HEATING, COOLING & VENTILATION

HEATING, COOLING & VENTILATION

Projects for Comfort and Value

Jay Hedden

A DIVISION OF FEDERAL MARKETING CORPOR
24 PARK WAY, UPPER SADDLE RIVER, NEW JE

CREATIVE HOMEOWNER PRESS®

Manufactured in United States of America

Current Printing (last digit)
10 9 8 7 6 5

Editor: Shirley M. Horowitz
Associate Editor: Gail N. Kummings
Art Director: Léone Lewensohn
Designers: Léone Lewensohn, Paul Sochacki
Illustrator: Norman Nuding

Cover photo courtesy of Velux-America Inc.

We wish to extend our thanks to the many designers, companies, and other contributors
who allowed us to use their materials and gave
us advice. Their names, addresses, and individual identifications of their contributions can
be found on *page 141*.

ISBN: 0-932944-40-X (paperback)
ISBN: 0-932944-39-6 (hardcover)
LC: 81-67294

CREATIVE HOMEOWNER PRESS®
BOOK SERIES
A DIVISION OF FEDERAL
MARKETING CORPORATION
WAY,
RIVER, NJ 07458

PROJECTS

In addition to basic techniques, projects can be found in the following pages.

CONTENTS

In certain parts of the country, particularly in the West and South, you can use a plastic solar pool blanket to cover and heat the pool. Individually sealed air bubbles cover the sheet, which keeps the pool temperature between 75 and 80 degrees F. under normal circumstances.

1
Conventional Heating Systems

TYPES OF HEATING SYSTEMS

The age of your house has much to do with the type of furnace you have. Houses 50 or more years old may have a gravity hot air system. This kind of furnace is the "octopus" that fills most of a basement with large round pipes that angle up through the floor. Other homes of the same era may have gravity hot water systems or even steam systems.

Homes built in the last several decades probably will have forced hot air systems. A small percentage will have hot water systems that pump the water through the pipes and radiators. These hot water (hydronic) systems were used in houses with concrete slab floors right after World War II, the water being circulated in pipes buried in the slab. While this is an efficient method of heat-

ing, lack of insulation around the edges of the concrete slab allowed much of the heat to be lost. We now know that insulating outside a slab, or even a foundation wall, causes the masonry or concrete to act as a "heat sink" that stores heat that is released into the house when the temperature of the air drops below that of the masonry.

No matter how old your heating plant is, if it is working well, don't replace it just to get a more modern system. The difference in efficiency may be so slight that you never will recover the cost of the new heating appliance. If, however, you cannot afford your current fuel costs, there are a number of improvements and conservations you can undertake that cost much less than a new heating system.

Gravity Hot Air System

This is basically a small metal box inside a larger one. A flue pipe passes through the bigger box and leads from the smaller box to the chimney. A fire in the smaller box heats the air in the space between it and the larger box; the air rises to the top of the furnace to a chamber called the "plenum." Hot air ducts run from the plenum up into the house. The lighter hot air moves up into the house through registers to heat the various rooms. As it cools, it becomes heavier and drops down to the floor. The cooler air moves through the return air ducts to the furnace. Here the air is heated again until it rises up to the rooms. This air circulation continues as long as there is a fire in the furnace; the fuel can be wood, coal, oil or gas.

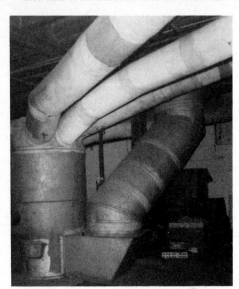

The gravity hot air furnace was often called an "octopus" for obvious reasons; its "arms" took up most of the basement. The larger duct to right is the cold air return; smaller ducts are for hot-air supply.

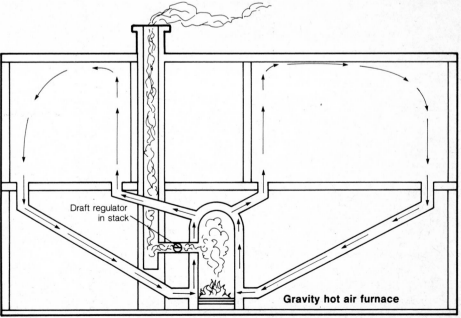

Draft regulator in stack

Gravity hot air furnace

The simplest of all hot air furnaces is the gravity unit that is quiet, but slow to respond to demand for higher temperature. Hot air ducts lead to registers on inside walls; cold air returns are on outside walls.

Advantages and Disadvantages

Because there are no moving parts in a gravity hot air system, it is wonderfully silent. One disadvantge of the gravity system is that the very small pressure difference between the hot and cold air will not allow it to move through a filter. Filters often are fitted over hot air registers to remove some of the dust, but they generally are cheesecloth or other coarse material that does not filter very efficiently. Because of the slow flow of the air, the hot air ducts are kept as short as possible. This means that in most cases the hot air registers will be located on inside walls of a house, closest to the furnace. The return air registers thus will be on outside walls. Warm air moves past the warmer inside walls first; it then drops down the cooler outside walls, windows and doors to the registers. On cold winter evenings the outer areas of some rooms will be quite chilly. Because of the slow movement of the heated air, the furnace is slow in responding to a higher setting on the thermostat.

An older house that is well built can be quite comfortable once the insulation has been upgraded. The expense of installing a newer type of furnace might not be worth the cost.

Forced Hot Air Systems

One of the main differences between this and the gravity model is that the cold air ducts of forced air heat run to a main return duct from each room. A blower driven by an electric motor pulls the cold air into the return ducts, blows it through the heat exchanger in the furnace and back into the hot air ducting.

Advantages and Disadvantages

Because the air moves quite rapidly, the furnace can be smaller with a hotter fire. It is thus much more efficient. Because the fan will blow the air through ducts in any direction, the furnace does not need to be centrally located. It can even be in the attic or a first floor utility room. In some setups the furnace will be in an attached garage. If the garage is not insulated, the jacket of the furnace must be insulated in order to maintain the efficiency.

Humidifiers

Because moist air carries more heat than dry air does, humidifiers often are

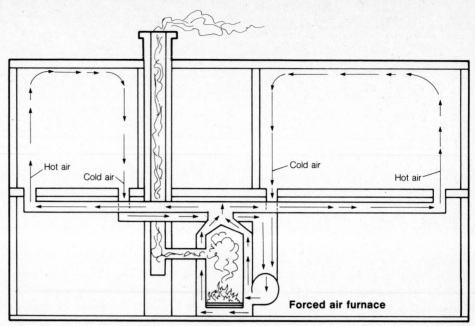

Forced air furnace

A forced air furnace has a blower that pulls air into the return ducts, pushes it through the furnace heat exchanger and then through the hot air supply ducts.

installed in a gravity- or forced-air heating system. These devices are automatic. Water is fed to them from a supply line by an electric solenoid valve that is triggered through the electrical circuit of the blower motor on forced air systems. For gravity systems, the humidifier has a float valve that allows more water to enter the unit as the water is evaporated by the passage of hot air through it.

Added advantages of moist air are that it does not dry out furniture, woodwork, and noses and throats of the householders (this dryness causes respiratory problems). Chapter 9 more completely describes Humidity Control.

Hot-Water Gravity Heating

Operation Hot water heating systems also can work on the gravity principle

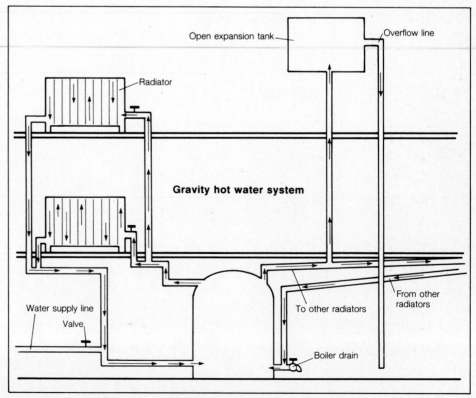

Gravity hot water system

Hot water gravity systems are the fluid counterparts of hot air gravity furnaces. They are efficient but slow to respond to demands for more heat.

just like the hot air setups. Hot water rises up through the boiler and into a main that angles upward around the basement walls. Smaller pipes extend up from the main to radiators in the rooms upstairs. The water flows through the radiators, then drops back through a line at the other end of each radiator to a return line that directs the cooled water back to the boiler. Here it is heated again to flow back to the radiators.

Advantages and Disadvantages

Water can carry an enormous amount of heat, much more than air, so a hot water system provides a more even heat, However, this system may be slow to respond to setting changes.

Adding Efficiency

One way to increase the efficiency of a gravity hot water system is to install a pump to circulate the water. If you have a gravity hot water system that is in good condition, but is slow to respond to a demand for a higher temperature in the house, installing a pump might be all that is necessary to solve the problem. There is no doubt that this addition would be a lot less expensive than installing a complete new furnace.

Installing the Pump

A do-it-yourself homeowner with a lot of plumbing and mechanical experience might be able to install the pump, but we recommend that a professional heating contractor do the job in most cases. For example, it is necessary to determine if the pump should run constantly or somehow be actuated only when the burner was ignited. This could require some rather sophisticated controls. The capacity of the pump also needs to be figured out, to match the volume of water in the system and to move it at a predetermined rate for the greatest efficiency. If the water moves too fast, heat in the radiators is not removed from it as rapidly as it should be. On the other hand, too slow a movement of the water allows the water to lose too much heat before it returns to the boiler. It would, at best, not be much better than a straight gravity system.

Hydronic (Forced Hot Water) Heating

The hot water version of a forced air system is called a "hydronic" system. Heated water is pumped through the system by one or more centrifugal pumps. Like the forced air system, it can have a smaller boiler, pipes can be smaller and the boiler can be located almost anywhere, even outside the house. There are three ways that hot water (or steam) usually is routed through the radiators: the "series loop", the "one-pipe" arrangement, and the "two-pipe" system.

The Series Loop This is the simplest arrangement. The hot water is pumped through the boiler into the main supply line that goes from one radiator to the next. The radiators are like finned sections of pipe; the water passes through them and the pipe to return to the boiler. The water is reheated and repeats its path through the pipes and radiators.

Although this system requires the fewest pipes and fittings of any arrangement, the disadvantage is that no single radiator can be shut off. If it is shut off, all circulation stops.

Because all radiators heat up and cool off at the same time, the series loop system is best suited to smaller homes that do not need variations in temperature from room to room.

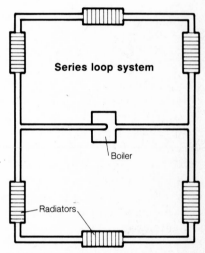

Series loop system

Boiler

Radiators

A "series loop" system for hot water system is simplest arrangement, but is not very flexible. If one radiator is shut off, all are shut off. This arrangement works satisfactorily only for a small house.

The One-pipe This second method of routing water or steam has, like the series loop, piping that leads from the top of the boiler to the radiators. The pipe then returns to the bottom of the boiler. The main difference is that a branch pipe from the hot water main runs to each radiator, while another pipe runs from the outlet of the radiator to the return main. The shutoff valve on the supply line at each radiator can be opened or closed to regulate the amount of hot water entering the radiator, or even to shut off the water completely. The water in the supply line simply flows past the shut off radiator to the next in line, then on to the other radiators in the system and back to the boiler.

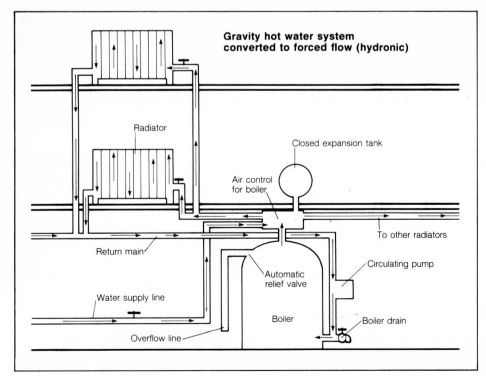

Gravity hot water system converted to forced flow (hydronic)

Radiator

Closed expansion tank

Air control for boiler

To other radiators

Return main

Circulating pump

Water supply line

Automatic relief valve

Boiler

Boiler drain

Overflow line

A simple way to increase the efficiency of a gravity hot water system is to install a pump to circulate water from the boiler through the radiators. This job, which calls for some skills and fairly sophisticated controls, is best left to a professional.

More pipe, fittings and labor are required for the one-pipe system, as opposed to the series loop system. However, it is popular because it offers individual control for each room.

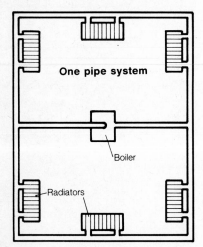

A "one pipe" system for hot water has supply and return lines from each radiator leading to a main that is both the supply and return line. In this system cool water is warmed by the hot, while the hot is cooled by the returning water.

The Two-pipe This resembles a gravity hot water system, with a pump that forces the water through the pipes and radiators. Like the gravity system, there is a separate supply and return main with branches to each radiator. Hot water from the supply main flows through each radiator, then discharges into the return line so the hot and cold water never mix.

Because the water is pumped through the two-pipe system, there is very little difference in the temperature of the first and last radiators, so this system is excellent for larger houses.

Zone Heating

In split-level and two-story homes, the heat tends to rise to the upper level through stairways. The upper level becomes warm while the lower level cools off. If the same amount of heat were supplied to both levels, the upper area quickly would become too warm. Therefore, larger homes—two-story, and some moderate-size split-level homes—are fitted with "zone heating." This can be either forced air or hydronic; its purpose is to direct heat to one zone or section of the house.

The house is divided into zones; each has its own thermostat. When the thermostat on the lower level signals for heat, but the one upstairs does not, heat

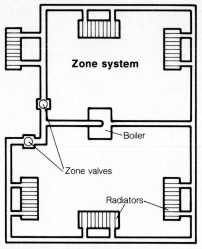

"Zone heating" can be either forced air or hydronic, with 2 or more valves or dampers electrically controlled by a separate thermostat. This system is good for split-level and multi-story houses, and for larger one-floor homes.

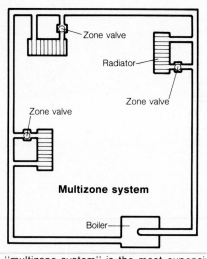

A "multizone system" is the most expensive to install, as there are controls for each zone with a matching thermostat. The hydronic system is shown, but a hot air setup would be the same, with dampers rather than valves for control.

is directed only to the lower level. In a forced air system, electrically operated shutters close off the ducts to the zone that does not require heat. Electrically operated valves do the same job in a hydronic system. Zone systems are quite flexible and heat can be supplied to all zones at once or to one or more in a multizone system.

Electrical Heating

There are several types of electrical heat systems: the antifreeze system, resistance coils, and localized electrical heaters. Electric heating usually calls for one or more baseboard units in each room.

The 240 volt wires for the heating system are easily run through the structure

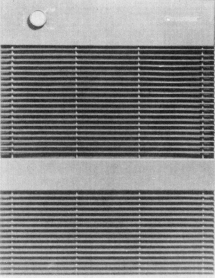

Electrical resistance heating is the most expensive to operate, but for room additions, basements and other rooms where the existing furnace does not supply adequate heat, it can be a workable solution.

of a house and the lightweight baseboard units are quickly connected. No flue or chimney is required with electric heating, and each unit in a room can be turned on or off, or set at various temperatures. It is the most flexible of all heating systems. There is no getting around the high cost of running such a system, however, which is why it is not very popular.

Resistance Coils This is the simplest electrical heat supply system. Its coils are much like those in a range top. The wire glows red hot, but it is covered with insulation as for the range-top units.

Antifreeze System Less expensive to power are individual baseboard units that contain a tank of antifreeze that is heated by the resistance wire. When the water/antifreeze solution reaches a preset temperature, a thermostat shuts off the electricity. Heat flows from the liquid until its temperature drops below the thermostat setting, when the heater again turns on. This kind of heater uses about half the electricity of a straight resistance heater, but the first cost is about twice that of a comparable straight resistance unit. In the long run, such a heater is more economical.

Local Electrical Heaters Other electrical heaters utilize the infrared spectrum to heat people, furniture, rugs and the like, but not the air. These heaters are used for "spot" heating, as in bathrooms (most motel bathrooms have them), but the units can even be disguised as pictures on a wall. Portable infrared heaters with quartz tubes also are available. These can be carried from room to room to keep a person warm, while the main thermostat in the house is set down to 60 or 65 degrees F.

Gas Furnaces

Most gas furnaces have a burner with a number of jets, as on a gas range. Some furnaces may have a single large jet of gas flame that is spread by a deflector plate to heat a surface. Both types will be lighted by a pilot light that also does another job; it heats a thermocouple that converts the heat energy to a tiny electric current. Although the current is small, it is sufficient to hold open an electrical valve that supplies gas to the furnace.

Keeping the Pilot Light Going If the pilot light goes out, the thermocouple cools, the electric current shuts off and the gas valve closes. Thus, even though the thermostat calls for heat and the burner valve opens, no gas is supplied to it. This prevents the gas from flowing unchecked into the furnace and the house and setting up a deadly, explosive condition in the house. The pilot light must be relighted, which requires holding open a special valve for a specified time. Relighting instructions are on or near the pilot light of the furnace (the same is true of gas hot water heaters). Modern furnaces now utilize an electric ignition system.

To prevent a gas pilot light from being blown out by downdrafts from the vent pipe and chimney, gas furnaces have a draft diverter either built into the furnace or fitted in the vent pipe that connects to the chimney. The most common

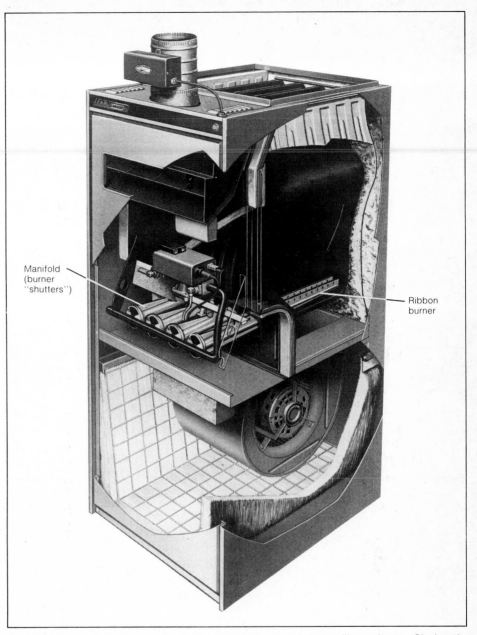

Manifold (burner "shutters")

Ribbon burner

To adjust the fuel/air mix, look for the burners and a round device like a shutter. Closing the shutter reduces the amount of air going to the burner; opening the shutter increases the air flow. Adjust the shutter so the flame is blue. A flame that has too little air will be yellowish. Opening the shutter a little at a time causes the yellow to disappear and more blue to show in the flame. An excess of air will cause the flame to hiss sputter and the flame will not produce the maximum of heat. To adjust the shutter, turn it only a little at a time. Aim for just below an excess of air—there will be a "soft" sound and the flame will be mostly blue.

type of diverter design is somewhat like one funnel inside another, with air space between. Although heated air normally flows up from the wide part of the funnel to the narrow section, if a draft blows down the chimney and reverses the flow, most of the air pressure dissipates through the space between the funnels and into the furnace room. This prevents the pilot light from blowing out. Under very severe wind conditions, however, a pilot light still can be blown out because not all the wind pressure is relieved through the draft diverter. This is another reason

why electric ignition is a more reliable system.

When to Turn Off the Pilot Light A gas pilot light burns an enormous amount of fuel during a year. Most homeowners leave the pilot light on the year around, having been told that it helped keep dampness out of the heating plant during the summer, thus minimizing rust problems. Not so, we now have learned. Gas burns to create hydrogen oxide, since hydrogen is the main burning component, and hydrogen oxide is water, H_2O. This warm water vapor condensed on the cool surfaces of the heat exchanger, usually cast iron, and caused rust. This means it pays both in savings on the cost of gas and the longer life of your furnace to shut off the pilot light during the summmer months.

Oil Furnaces

These employ a "gun" through which oil and air are pumped into the furnace combustion chamber. The pump that forces oil to the spray nozzle of the gun and the blower that supplies the air for the burning are powered by one electric motor. Rather than a pilot light, as with a gas furnace, an oil burner is ignited by a high-voltage spark across a pair of electrodes positioned directly in the stream of oil and air.

Burner Control Some burners have sparks that are maintained during the burning cycle, which means that the electrodes need adjusting at the beginning of each heating season. Their spacing must be correct for the spark to be generated. At some point in the life of an oil burner, the electrodes must be replaced.

If for some reason the air/oil mixture fails to ignite, the burner will quickly shut off so oil does not continue to spray into the combustion chamber. Automatic burner shutoff is handled by one of two devices. The most common is a stack control located in the metal pipe leading to the chimney. This is a bimetal device, in which one strip of metal expands more rapidly than the other when heated and the strip bends. When the strip cools, it straightens, and when cooled still further it will curve in the opposite direction. This bending reaction to heat and cold is utilized to turn a switch on or off. In the case of an oil burner, when the control is cooled it shuts off the electricity to the pump/blower motor and to the ignition electrodes.

Photoelectric Devices On newer oil burning furnaces the automatic shutoff is handled by a photoelectric device. This unit resembles the cells that automatically turn on yard lights when it gets dark. However, in the case of a furnace, the cell is used to do the opposite job; the cell shuts off the electricity. The cell is aimed at the spot where the flame should be when the air/oil mixture ignites. When the flame is normal, the photoelectric cell provides electric current to keep the power switch to the pump/blower motor closed. If no flame appears, the photocell's electric current diminishes and the power switch opens to shut off the motor.

Other Safety Devices Whether based on water, gas or oil, furnaces do need safety devices. One such mechanism is the high-limit switch in the plenum of a hot air furnace or in the supply line of a hydronic system. If the air or water reaches a higher temperture than that for which the heating plant is designed, the switch will shut off the gas or the current to the pump/blower motor. When the temperature drops to a safe level, the switch closes and current is again supplied to the oil/blower motor of an oil furnace. In the case of a gas furnace it usually is necessary to relight the pilot.

Because a high temperature indicates some kind of malfunction, you should immediately call a serviceman. Furnaces are not something that the average homeowner should tinker with, because such attempts might override a safety device and cause a potentially hazardous condition.

REDUCING FUEL CONSUMPTION

In the past, the heating efficiency of an average gas or oil furnace was considered to be in the 60 to 85 percent range. When the energy shortage struck, more accurate testing was done and the actual percentages turned out to be closer to 35 to 50 percent. This means that for every dollar you spend on fuel, only 35 to 50 cents is converted to heat.

Increasing Furnace Efficiency

Modern high-efficiency furnaces, with electronic ignition and precisely engineered burning mechanisms, now really do reach the 75% efficiency range.

For even higher efficiencies there are "flame retention" burners that can be installed in an oil burning furnace to increase the heating efficiency to 85 percent. The Sloan Valve "INTERburner" is one such device. It takes in less air than a standard burner, but it mixes the air and oil for more complete burning. The Blueray heating system recirculates unburned gases for more thorough combustion and higher burning efficiency. Replacing your present burner with one of these high-efficiency units can cut fuel usage and costs.

An oil-fired furnace requires the services of a professional. The boiler here is being cleaned with wire brush and high powered vacuum. Soot that collects on the heat exchanger acts as insulation and reduces efficiency.

HOW TO KEEP A GAS FURNACE RUNNING WELL

Although a gas furnace seldom needs cleaning, check that the fuel/air mixture is not too rich.

Replacing Rusted-Out-Pipe Inspection of the metal stack may show that condensation has rusted out the underside, thus reducing the draft and furnace efficiency. It also will allow toxic combustion products to enter the basement. If the stack pipe has rusted out, new sections can be purchased at most hardware stores for a simple do-it-yourself replacement.

Cleaning or Replacing the Dust Filter A regular maintenance job for any furnace is cleaning or replacing the dust filter. This alone can save up to 10 percent of your heating cost. During very cold weather, or when the central air conditioner is running, clean or replace the filter more often than the ordinary schedule. In these periods the furnace fan runs almost constantly and the volume of air being circulated can be up to twice as much as during more moderate weather periods.

Lubricating the Bearings On some furnaces the fan and motor bearings require a few drops of oil each season. Do not overdo this lubrication; if the instructions say two or three drops of oil, do not apply four or five. The excess oil can run down into the electric motor and cause problems, or can run through the fan bearings to collect dust that reduces the efficiency of the fan.

Dust Removal If the fan is accessible, brush or vacuum the separate blades to remove dust. It also is a good idea to vacuum dust and debris from inside a furnace cabinet. This will reduce the dust that can be picked up in the air stream and blown around the house.

A gas furnace burns cleanly enough so that it needs service only every three to five years unless there is a specific malfunction. The gun of an oil furnace and nozzle should be cleaned each season, and the electrodes should be cleaned, adjusted or replaced as required. A special heavy duty vacuum cleaner can be used to clean the firebox and inside of an oil furnace, but an ordinary shop vacuum cleaner should not be used. The oil fumes can collect inside the vacuum and present a fire hazard.

Insulating Ducts You can save heat by wrapping hot air ducts with insulation made especially for ducts. If the ducts run through an unheated crawl space, they absolutely must be insulated. The energy you save will be considerable. If you don't insulate the ducts, check the duct joints to make sure they are not leaking air. To stop leaks, wrap the joints with duct tape. The biggest air loss comes from elbow joints where a duct makes a turn.

Duct Fasteners You should also fasten down, pad, and replace faulty duct hangers. This will stop noise and vibration.

Insulate heating ducts in cold crawl spaces; tape the joints with duct tape.

Before vacuuming the furnace, turn it off and let it cool. Clean burner chambers, blower housing, motors, valves, and thermostats.

Clean furnace vents annually. If furnace is in a room with a louvered door, vacuum the louvers to aid air flow.

Check and replace damaged duct hangers. Pad noisy ducts by sandwiching insulation between duct and hanger.

CORRECTING THE SLOPE OF THE CHIMNEY PIPE

Even as simple a problem as not enough slope on the pipe from the furnace to the chimney can reduce the chimney updraft and lower the furnace efficiency. Raising the chimney end of the stack will produce the correct upward angle of the stack to create a good draft. The stack will then slope away from the chimney.

Creating an Opening Shut off the furnace and let the pipe cool. Remove the sections of pipe. Examine them to see if sheet metal screws or blind rivets have been used to join the sections; remove them as necessary. Break away the mortar around the opening into the chimney.

The base of a chimney in older homes (the portion in the basement into which the furnace and hot water tank stacks are connected) might be brick. In newer homes it probably will be concrete block. To raise the chimney end of a stack you may have to remove a brick or break a section of concrete block. If you have no experience with masonry, it would be better to call in a mason to do the job. For an experienced do-it-yourself homeowner, removing one brick or breaking out a portion of one block should not be a problem because it will not cause structural damage to the chimney.

Repositioning the Stack End Fit the end of the stack into this opening, making sure it projects no more than about two inches into the chimney opening. Then block it in place with scraps of block or brick. Replace the metal stack between the chimney and furnace; support it on blocks as needed. This work will have to be done with the furnace and hot water tank shut off, of course.

Mortaring the opening To plug the opening under or around the stack, and to reinforce the mortar, you can stuff in some metal wire screening or hardware cloth. Get a bag of ready-mix mortar from a hardware store or home center and mix the dry ingredients thoroughly. Pour all but two or three pounds back in the bag. Mix the dry powder with water to make a thick paste. The mortar is a standard mix of sand, portland cement and lime. It need not be firebrick mortar because it is not exposed to open flame, as would be the case in a fireplace or a furnace firebox.

Before applying the mortar to the opening in the chimney, thoroughly wet down the opening. This is to prevent suction—the pulling out of the water in the mortar by the dry brick or concrete block. Apply the mortar to the opening, packing it firmly in place with a small trowel. If the mortar tends to slide down in the opening, it's too wet. Add a bit more of the dry powder to thicken the mix, stirring it thoroughly. Apply it again and force it into the screen or hardware cloth; bring it flush with the brick or block. Pack it firmly around the stack (or stacks, if the hot water tank has a separate stack going to the chimney; in most cases, however, the stack from the hot water tank will be smaller than the furnace stack and will be connected to it at a point near the chimney).

Let the mortar set properly before turning on the furnace; wait several hours. Heat will dry the mortar too quickly.

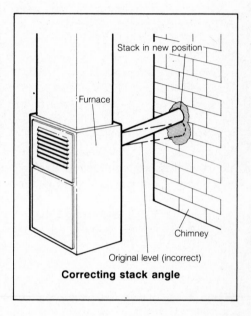

Stack in new position

Furnace

Chimney

Original level (incorrect)

Correcting stack angle

ADDING A PROGRAMMABLE THERMOSTAT

Most homeowners today are concerned with rising heating and cooling costs. One helpful and money-saving project, which is easy to add to your home, is a programmable thermostat. This can be set to adjust the temperatures in your home automatically for waking and sleeping hours. The thermostat also can include indicators that set temperatures for those hours when the house is usually occupied and for those hours when it is not. Since your heating-cooling system will not use as much energy when the house is empty, you will reduce fuel usage and cut costs.

Wall Thermostat Operation

Wall thermostats are used to control furnaces. They have bimetal devices that respond to heat by bending and closing an electrical switch. Most modern thermostats operate at very low voltage and amperage and will be rated in "milliamps." The very small current is sent through small wires to actuate an electric solenoid switch. This switch closes and carries the full 120 volt household current at the furnace to power the blower or pump.

Modern thermostats utilize a tiny electric heater to help keep the temperature in a home at a more constant level. When room temperature drops to the setting on the thermostat, the furnace burner comes on to heat the air or water in the heat exchanger. It takes a minute or two for the temperature in the furnace to reach the desired level and, in the meantime, the temperature in the room drops still further below the thermostat setting. Finally, heat begins to reach the room and it warms the bimetal in the thermostat until it warms enough to shut off the furnace. Because even the best thermostat takes a little time to respond to temperature change, the room gets warmer than the setting on the thermostat. This thermostat lag causes a noticeable rise and fall of temperature in the house. To prevent this fluctuation of temperatures above and below the thermostat setting, a very small heater in the thermostat turns on as soon as the burner is ignited. This heater warms the bimetal so it begins to bend toward the switch shutoff point before the temperature in the room starts to rise and affect the thermostat. The thermostat thus shuts off the burner before the room reaches the desired temperature. Because there still is heat in the system, the blower or pump continues to work, adding enough heat to the room to come close to the set temperature. If you notice wide fluctuations of temperature in your home, it is quite possible that the heater in the thermostat is not working. The thermostat must be replaced; it is not economical to repair these devices.

Types of Thermostat Units

The units come in a variety of styles and are produced by several manufacturers. There are some units that replace heat-only thermostats, which have two or three wires. Other units will replace thermostats that are combination controllers for heating and air conditioning. The latter may have four or five wires. Since energy-saving, programmable thermostats vary considerably from one brand to another, be sure to get full instructions with the unit when you buy it.

Components All the models come with a wall plate that is fastened in place over an opening through which the existing thermostat wires run. On top of the wallplate, you attach a subbase that is wired to be compatible with your thermostat. The actual programmable thermostat then fits over the subbase. Contacts on the back side of the new thermostat touch contacts on the front of the subbase to complete the necessary circuits.

Installation Guidelines

The unit described here is made by Honeywell. One major requirement for the thermostat is that it be exactly level, because the mercury switches used as controls must be level to operate properly. A mercury switch is a glass bulb inside which there is a small pool of mercury and two wire conductors. When the bulb is tipped, the mercury runs down to immerse the conductors. Electric current then passes through the mercury from one conductor to another. When the bulb (switch) tips back, the mercury moves away so the conductors are separated and the electric current stops. ("Silent" light switches work on the same principle. The switch toggle shifts a small container of mercury up and down to control the light.) To insure a level thermostat, use a small spirit level.

Information You Must Know

Before you buy the new thermostat, you need certain information about the existing one in order to purchase a programmable thermostat that will properly replace your existing one. Note the brand name and model number of the existing thermostat. Then remove the cover so you can determine how many wires are in the thermostat connection. If the thermostat controls heat only, there will be two or three wires; if it controls both heating and air conditioning, there may be four or five.

Check also to see if you have the instructions that came with the existing thermostat. They will help when you wire the new programmable unit. If you do not (and most of us will not, since the previous owner or builder will have discarded them), installation instructions for the new unit usually include "typical" wiring diagrams.

Installation Procedures

We will repeat: programmable thermostats vary from one brand to another, so be sure to get complete instructions with your unit. However, the following is a general breakdown of the steps involved.

Removing the Old Thermostat When you install the new thermostat, be sure that the electric power is off. To accomplish this, remove the fuse or pull the circuit breaker that runs your furnace fan. Lift off the cover plate from the old thermostat. Carefully note which wires go to which connection in the existing thermostat. The connections will be numbered or lettered, and comparable identification marks will be on the connections of the new programmable thermostat. The easiest way to keep track is to tag each of the lines.

Inspect the condition of the wires carefully. If they are discolored, or if the insulation is cracked or broken, you must either cut back the wires to solid material, or you must replace the wiring leading from the furnace fan to the thermostat.

Now remove the old thermostat base from the wall. Plug the wall opening through which the wires come. This is important. Otherwise, a warm or cold draft through the opening could affect the operation of the thermostat.

Installing the New Unit The Honeywell thermostat pictured here has a built-in clock. This, coupled with the programming, governs the heating and cooling cycle. The clock is run by a small battery, which must be charged before

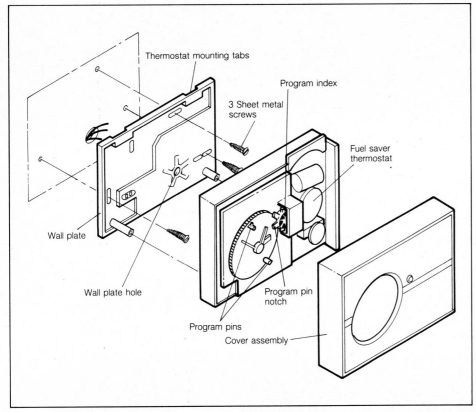

Thermostat mounting tabs

Program index

3 Sheet metal screws

Fuel saver thermostat

Wall plate

Wall plate hole

Program pin notch

Program pins

Cover assembly

Pull the wires through the wall plate. Position the plate and place the mechanism. The program pins in the clock wheel turn the system on and off at preset times.

the thermostat can function. From then on the battery recharges while the thermostat is working. Methods for acquiring the initial charge will vary according to manufacturer; see the instruction sheet for your model.

Making Wiring Connections Draw the wires through the opening in the wallplate and fasten the wall plate loosely in place with the screws provided. Level the plate and secure it firmly, but do not overtighten the screws or the wall plate threads will be stripped. Fasten the wires as shown.

Then tuck the wires back into the hole—without dislodging the patching material there. Protruding wires can catch on the mechanisms in the thermostat and cause it to malfunction.

Attaching the Base At the back of the thermostat base are small spring fingers. These match up with terminals on the wallplate you just positioned. Install the base carefully so you do not damage the springs.

Setting the Heat Anticipator Check the primary control or gas valve of your heating system to find out how much current it draws when it is operating. This is your heat-anticipator rate. On the lower right side of the base is a dial. Set this to match the rating you found.

If you cannot find a rating, check with the instruction sheet of your unit to find how you can set this dial.

Programming the Thermostat The programming for the Honeywell unit is handled by three interrelated functional components. The first is the built-in clock. This is set up in two 12-hour sections, one light-colored for daylight hours, the other dark-colored for night-

time hours. The second component consists of two temperature levers, one of which is blue for the lowest temperature level, and one which is red for the highest temperature level. The third component is a set of pins that fits into cogs built into the clock wheel. These send a signal to the control mechanism, indicating that either the red or blue temperature level should govern the system. The pins are color-coded red or blue to indicate which level is being signalled.

Keep in mind that it will take approximately one-half hour for the temperature to change from one level to another. (For this reason, two pins cannot be closer to each other than one hour.) Therefore, if you set the night temperature for your furnace at 60° and

you want that to switch to 68° F. by 7 o'clock in the morning, the red pin would be set at 6:30.

Multiple settings The thermostat can handle more than two pins. If the house is empty from 8:30 to 4:30 every day, set a blue pin at 8:30 and a red one at 4:00. In that way you can have a warm house in the morning and in the evening, while you are there, and a cool one while everyone is gone or asleep.

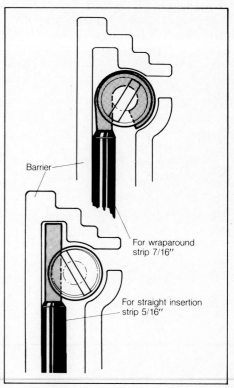

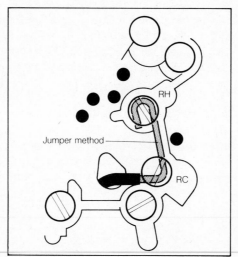

If your system has a single transformer for heating and cooling, strip enough insulation from the wire to connect terminal RC and RH.

Wires can be fastened to the terminal screws in two ways. Either curl them around the terminal post or insert them beside the post.

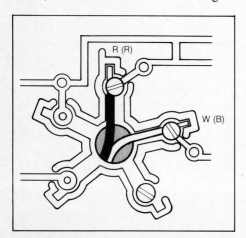

The wires lie within the design ridge. This arrangement will prevent interference with the thermostat mechanisms.

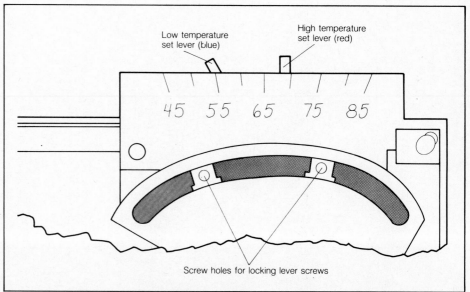

Built into the temperature control are two color-coded levers. The blue lever sets the cool temperature; the red lever sets the warm temperature. If desired, lock the levers in place.

OIL TO GAS CONVERSION

Converting an oil burning furnace to gas has become popular in the last couple of years because gas is much cheaper and in some parts of the country oil has been in short supply. With the deregulation of oil prices for domestic producers the cost of oil will no doubt continue to increase. Gas also has been deregulated, however, and forecasts for the price of gas indicate that it soon will be equal to that of oil. Therefore, unless oil has really become in short supply in your area, and gas has become more available, it would not seem to be economically feasible or practical to change from oil to gas.

Several factors are involved also; a gas conversion unit for an oil furnace may not be very efficient. Replacing an oil furnace with a complete new gas unit can cost around $2000. On top of this, a utility company may require that you pay for the pipe line from the street to your house. This additional cost must be included in any changeover from oil to gas.

ADDING A NEW HEATING AND DUCTING SYSTEM

Do not accept any one judgment that your furnace is dangerous and needs to be replaced immediately. Get a second opinion. If you are told that your furnace needs to be larger to supply adequate heat to your house, first get an energy audit.

Planning the Ductwork Planning the system is the real key to success. Work out a scale drawing of your house and then visit Sears, Roebuck or some other furnace producers to design a heating plan. There will be a fee for this service, but it will be worth it because you will be assured that the heating plant operates properly. In addition, a good plan will save money because it will use the least amount of ducting and connectors necessary.

The economy-duct system For a small house, an economy duct system can be used. This has the furnace on the first floor of the house and the hot air supply ducts routed through a dropped section of the ceiling. The cold air inlet is located in the furnace itself. A main duct runs the length of the house, with branches off this duct to the rooms.

The radial system Similar to the economy system is the radial system, which has the hot air ducts in the attic running to the separate rooms. Again, the cold air return is in the furnace.

Where the furnace is to be located in the crawl space, the ducting is run under the floor from each end of the horizontal unit, but the hot air is supplied from one end while cold air is ducted to the other.

In a basement radial duct system the individual ducts run from the furnace plenum to the various rooms to supply hot air. Much more ducting is required for this kind of system, but it could be made a multizone system with control for each room. While the drawing shows the ducting to be at right angles and square to the furnace, it also could spread out in spiderweb fashion for shorter and more direct routing.

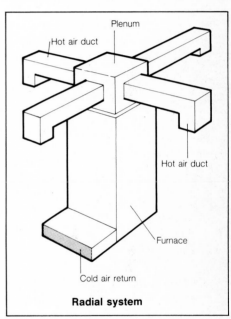

Radial system

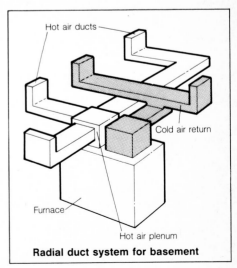

Radial duct system for basement

A radial duct system for a basement has individual ducts running from the furnace to each room, which makes this a fairly complex and expensive installation. Cold air returns also can fan out so each room has its own air return.

Economy duct system

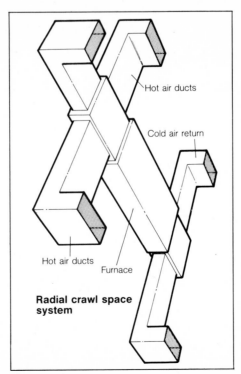

The economy duct system utilizes one large supply duct with branches to rooms; cold air return is one unit at the bottom of furnace.

In a radial crawl space system the hot air furnace is positioned horizontally; hot air discharges from one end to the other.

Radial crawl space system

The extended plenum system This setup is installed in a basement. The furnace is positioned vertically, but the hot and cold air plenums are both at the top of the furnace. This is a more expensive and complicated system to install, but very flexible and efficient. Although only two cold air ducts are shown below, several actually could be installed to route the air from various parts of the house. A main duct is used for the hot air, with smaller ducts leading from it to the various rooms.

Ducting variations There are a number of other variations of these ducting systems; some systems combine elements of two or more of the systems shown. It would probably pay to hire a heating engineer to design your duct system for you. You can then do some or all of the installation yourself. You will have a more efficient heating system and a comfortable house, and the fee will be paid back in a short time by the lower cost in your heating bills.

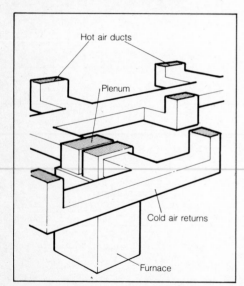

An extended phenum system is a large supply duct extended from the furnace plenum, with branches to various rooms.

Installing or Extending the Ductwork You will learn much of what you need to know about installing ductwork by studying your existing ductwork. If you do not have ductwork, or you think the ductwork you have is substandard, go to a subdivision where there is new construction taking place and for a few hours watch how the builders install ductwork. Round ducts, or duct "pipes" are frequently used in residential construction, but you may have rectangular ducts with which you must deal. The round ducts are easiest to work with and may be used for extensions off of both existing round ducts and rectangular ducts.

Cold air returns are created by nailing sheet metal to the bottoms of floor joists, when the ducts run parallel to the joists. Where the cold air ducts run at an angle to the joists, wide, flat ducting is used to assure maximum headroom in the basement.

Wide flat ducting also is used for the hot air ducts when they must run below the floor joists. Where the ducts can run between the joists, round ducting often is used. Adapters from round to rectangular are required for the transition, and a variety of such adapters and fittings are available to route hot and cold air to registers in the floors and walls of a home.

Physical characteristics of ducts and collars One end of each round duct section is crimped and one end is plain. The crimped end tapers down several inches so that it fits into the plain ends. Round duct typically is shipped open; that is, the duct has not been fastened together lengthwise. The pipe has a tongue and slot to lock the two pieces together. The fit is obvious when you see the duct. Thread the tongue into the slot. There will be a "click" when the two pieces fit properly. Do not hammer the tongue and slot edge together.

The collar, the section of duct that connects a new duct run to the furnace, is best installed by the subcontractor, or by you under the subcontractor's instructions. There are typically two kinds of collars: a straight collar and a take-off collar. The straight collar fits directly to the furnace plenum. A take-off collar is used where there is an extended plenum from the furnace; the take-off collar fits on the top or side of the extended plenum.

Cutting and Supporting the Ducts It will be necessary to cut the ducts; usually they can be cut with metal shears. Cut the ducts at the plain ends, not the crimped ends. Ducts may be

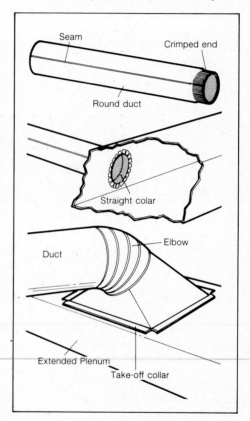

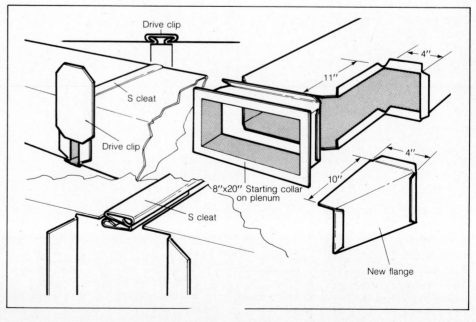

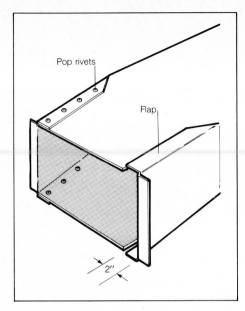

bought in sections as short as 2 feet, but the typical sections are several times that length. Each length of duct should be supported with at least one metal hanger.

The metal hangers, called "straps," look like metal tape with holes in them. Loop the metal hangers under the duct and bolt the hangers together at the top.

Nail through the holes into some structural member of the house such as ceiling joists or floor joists.

In no case should the ducts have less than one hanger every 10 feet. However, this rule does not consider the joints, which are weak spots. Tight joints are important to efficient, quiet operation of the system. So it is recommended that you install a hanger at each joint in addition to the hanger for every ten feet or less along the run of the duct. Most contractors do not do this. But since it is your house and the hangers are cheap—and it does not take much time to take this extra precaution—do it. This will add to the longevity of your system.

Installing the Ducts Pull the insulation away from the existing duct. Locate a joint and take the duct apart there. Install the proper length(s) extension by sliding the crimped end of the pipe into the plain end. Note that the crimped end of the existing duct points away from the furnace. This keeps the forced air from escaping the joint. You will be installing the plain end of your new duct to the crimped end of the existing duct.

Fit the plain end of the duct over the crimped end so that the plain end slides over the crimped end as far as it will go. Save the joint screws you removed when taking the duct apart, and drill holes to accommodate the sheet-metal screws you removed. Secure the joint with the sheet-metal screws and run some duct tape around the joint, covering the joint and the screws.

Insulation When you finish extending or adding all the ductwork, insulate the ducts and seal joints with a joint vapor barrier on the outside of the insulation (insulation and joint sealers are available at any heating/cooling supply house).

Dampers Locations of dampers, the metal flaps inside that control the amount of air that flows, should be planned by a professional. You can buy and install prefabricated dampers yourself. They come in 2-foot duct sections, already mounted within the ducts. You simply install the short duct section just as for any other run of duct.

HOW TO EXTEND DUCTWORK THROUGH AN ADJOINING WALL

The major problem is to make sure the wall cuts match on each side of the wall; that is, every point on the inside cut should face exactly opposite the same points on the other side of the wall. Always try to plan the duct route between wall studs, so you won't have to cut them. Always turn off the electrical circuit in the wall to be cut.

How to Cut a Duct Hole Through the Inside Wall Set a piece of cardboard over the open end of the duct to be extended through the wall and draw its outline on the cardboard. Set the drawing (template) against the wall where you want the duct to go through, and mark through each corner of the template with your pencil, identifying the corners on the wall. You could mark the corners with nails, if that is more convenient. If your wall is insulated, pull the insulation away from the studs before you mark the template corners.

With the markings for the duct corners (or several points around the circumference, if the duct is a pipe) transferred onto the existing wall, connect the dots between the corners (or round out the circle) with your pencil.

On a wall that is covered with Sheet-

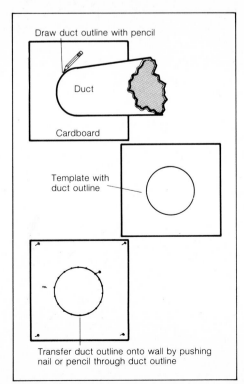

Draw duct outline with pencil

Duct

Cardboard

Template with duct outline

Transfer duct outline onto wall by pushing nail or pencil through duct outline

rock or plaster, draw the duct on it with the template, as described above. Then score the duct outline with a nail or similar pointed tool and remove the Sheetrock or plaster with a chisel, tapping it with a hammer along the scored lines and maintaining a smooth edge.

Cutting Wall Studs Studs always are connected at the top of the wall (to the wall plate) and to the bottom of the wall (sill). When you cut through a stud, you substantially weaken the stud and its original function as a wall member. To rectify this problem you need to add blocking at the top and bottom of the hole and tie it (toenail) to the uncut studs on each side of the hole. The cut studs are toenailed to the blocking. Ducts are not heavy; add a horizontal 2x4 at the top and bottom of the hole.

Dust protection When making a duct opening cut from the outside, you need to cover the inside hole to avoid brick dust or other debris being scattered around the house. Quarter-inch plywood, masonite, or similar materials temporarily nailed over the inside hole with light finish nails will serve.

CUTTING HOLES IN WALLS FOR REGISTERS AND GRILLES

You may use the same template procedure discussed above to cut holes for supply registers and return grilles.

The Hot Air Duct

The hot air duct should run to a register in the outside wall, or in the floor near the outside wall. Although the duct leading to the register can be round, the register in the wall or floor is rectangular, so an adapter called a "register boot" is connected to the round duct where it runs up through the floor or wall.

The Cold Air Duct

The cold air duct need not be completely metal (depending on local building

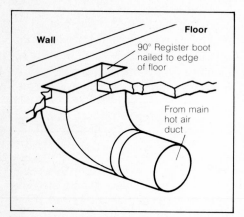

The easiest register to install uses a 90° register boot that comes up through the floor near the outside wall. A round duct connects the boot to the main hot air supply duct.

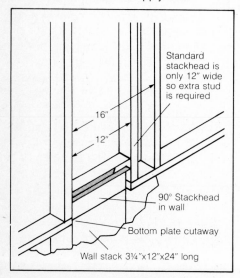

If the hot air duct runs up through the wall, a 90° stackhead directs air into the room. The wall duct runs down from the stackhead, to the duct that runs back horizontally to the main hot-air-supply duct.

codes). In some installations the floor joists in the basement or crawl space form the sides of the "duct," while the bottom is a piece of sheet metal. The

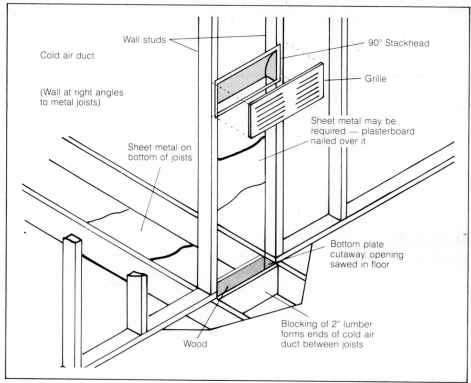

A cold air duct can be run up through wall after covering the bottoms of floor joists with sheet metal to create a duct. Use short lengths of the same stock as the floor joists to block ends of the duct. Depending on local code, you may have to run metal ducts up through wall. If not, use 90° stackhead to direct air into duct from room.

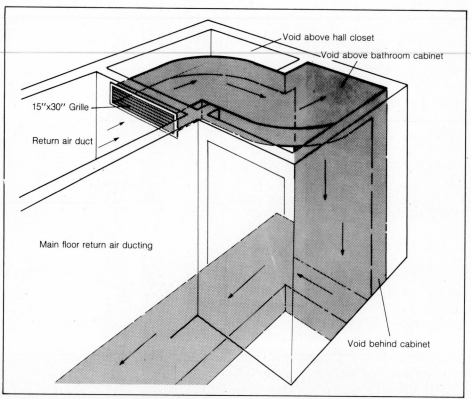

A return air duct, which must run alongside the main supply duct until it can be connected to an opening in the floor, can be hooked up to a void above and behind a floor-to-ceiling cabinet.

ends of the cold air duct are made by simply nailing short pieces of 2-inch lumber between the joists, one at each end. An adapter of some type, plus ducting, is used to join the joist section to the main cold air return duct.

In modern installations, the cold air register is fairly high on the wall. This is done for two reasons: (1) the warm air rises and is pulled into the return air duct to assure a complete circulation of air; (2) when central air conditioning is installed, the cooled air is discharged into the room through what is normally the hot air register. This cold air tends to stay low in the room, but gradually builds up so that warm air in the upper portion of a room is pushed up and toward the return air ducts. If the cold air register also were low on the wall, or in the floor, the cool air would simply move across the room close to the floor and return to the furnace, without affecting the warm air in the upper part of the room. The cold air return should be on an inside wall, which means that whenever possible you should arrange the wall studs to meet the floor joists. To do this, the bottom plate must be cut away, and an opening made in the floor.

If the wall you build is at right angles to the floor joists, cutting the opening presents few problems. If the wall is parallel to the floor joists, use round ducts and fittings below the joists. In a basement, where headroom can be a problem, choose shallow rectangular ducting and run it below the joists. In a crawl space that is unheated, insulate both the hot air and cold air ducts.

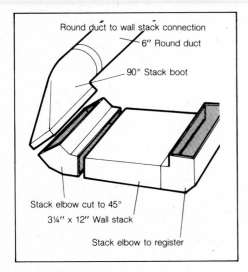

Round duct to wall stack connection
6″ Round duct
90° Stack boot
Stack elbow cut to 45°
3¼″ x 12″ Wall stack
Stack elbow to register

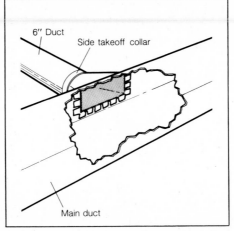

6″ Duct
Side takeoff collar
Main duct

Use a "takeoff collar" to connect round duct to a main duct. Cut a hole in the main duct; insert and secure it by bending edges over.

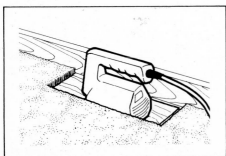

Before installing new registers, cut away any carpeting. Then saw out the opening.

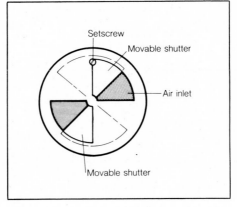

Setscrew
Movable shutter
Air inlet
Movable shutter

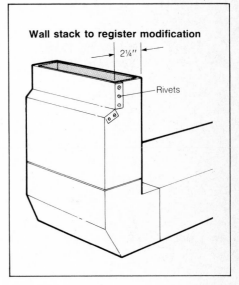

Wall stack to register modification

2¼″
Rivets

2
Air Conditioning Systems

SELECTING AN AIR CONDITIONER

The primary purpose of air conditioning is to cool the house. This can be achieved with a central air conditioning system, with individual or room units, or with evaporative coolers.

Central Air Conditioning Systems

The most effective air-conditioning method adds an air conditioning unit to an existing central forced air heating system. Ready-to-install air conditioning kits are available. With hydronic heating systems it is necessary to add ducting to direct the cooled air to the various parts of the house. Some air conditioning units can be installed in the attic with short runs of ducting down through the ceiling; this is somewhat less expensive than locating the unit in the basement and running ducting through the walls and under the floor.

Room Air Conditioning Units

Room air conditioners are cheaper to purchase and easier to install. A single unit can cool some small homes. Two or more units, in combination with fans to direct the air, can cool larger homes. They are not as efficient as a central system, but can be used effectively for cooling one or two rooms that can be sealed off from the rest of the house. This method may prove to be the most cost effective if cooling the whole house is too expensive. Small room air conditioners can be used to augment central units in rooms that are hard to cool.

Evaporative Coolers

Evaporative air conditioners or coolers are even cheaper, but are effective only in hot, dry climates. The humidity they

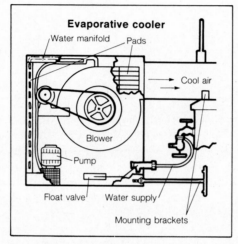

Evaporative cooler

Water manifold — Pads — Cool air — Blower — Pump — Float valve — Water supply — Mounting brackets

In an evaporative cooler, water enters the reservoir through a float-operated valve. A pump moves the water through tubes that distribute it to fiber pads. A blower forces air through the pads, where it is cooled by evaporation of the water. Excess water flows through the pads into the reservoir.

add to the air can be accepted in Tucson, Arizona but not in St. Louis, Missouri.

Reducing Air Conditioning Loads

When energy was cheap it was a simple matter to set the thermostat to a comfortable level and let the air conditioner work as necessary to cool the house. But there are many other ways you can help cool the house, thus reducing the load on the air conditioner and lowering its operating cost. Adequate attic insulation, along with good ventilation using vents or fans, can cut operating expenses by up to 30 percent by exhausting the hot air that builds up in the attic. This prevents leakage of the hot air to the rooms below (also see Chapter 3).

Hot, humid air can be kept out of the house by sealing openings with weatherstripping, storm windows and doors. Shading is another effective means of

lowering the home's internal temperature. You can use blinds, curtains, old-fashioned awnings, overhangs or "eyebrows" that keep the sun from entering directly through the windows and getting trapped inside. Proper landscaping can help by providing shading and directing cooling breezes.

Changes in daily household routines will also reduce the cooling load. Open the windows or use an attic exhaust fan after sunset to draw in cooler, outside air. Close windows or draw blinds to keep the daytime heat out. Use sparingly those appliances that create heat, especially during the hot part of the day. Close off those rooms where heat and humidity are high, such as bathrooms, and use vent fans to remove the heat. Be sure the clothes dryer is vented outside.

Comparing Energy Efficiency Ratings

If you are purchasing a new refrigerated air conditioning system, buy one that provides the necessary BTU's but also one with the highest EER (Energy Efficiency Rating). The higher the EER number, the less electricity the unit will use to cool the same amount of air. Consider possible future energy savings when deciding how much to spend on the air conditioning unit. A more costly unit may save enough money to pay for it by the next summer. Typical EER's range from 4 to 12. A unit with a EER of 4 will cost about three times as much to operate as one with an EER of 12. However, merely replacing an operating unit with one that has a higher EER rating may not be economical unless the savings will be substantial.

Carrier "High Efficiency" air conditioner (the "Round One") is a split system arrangement; the condensing unit is shown here. The evaporator installs in the furnace plenum.

Coleman D.E.S.® "Deluxe Energy Saver" split-system air conditioner has an EER rating of 13.2.

INSTALLING A ROOM AIR CONDITIONER

Room air conditioners are meant to be installed by their owners. They are often called "portable" but, except in the case of the smallest, they can weigh up to 300 pounds. This type of unit requires more than one person to carry it around and lift it into the window or other opening. The smaller units can be plugged into any standard 120 volt wall outlet. The larger, more powerful ones require their own 240 volt branch circuit. Unless a window unit was previously installed or you have an open circuit, you probably will need to have this circuit added.

Mounting a Window Unit Instructions will usually be provided with the air conditioner, detailing the procedures for correct installation. The first step is to prepare the cabinet for mounting in the window.

A small air conditioner mounted in a double hung window is held in place by the angles on the top and bottom of the cabinet. Mounting angles may already be attached to the cabinet when you buy it. If not, press a strip of adhesive-backed foam over the predrilled holes and attach the angles. There should be an angle on the top and bottom of the cabinet, and possibly on the sides, depending on the design of the unit.

Small Units The lower angle fits on the outside of the window sill, while the upper angle fits on the inside of the window sash. The weight of the unit holds itself in position. Side "wings" or a filler board is provided to fill and seal the opening on either side of the cabinet. The side "wings" are usually of a flexible plastic fabric and operate like a miniature folding door. They often seal poorly. A more effective seal can be had

by placing a filler board on the outside of the wings. Insulation placed between them offers an even better seal. (See "Permanent Installations", below).

To keep the lower window from being raised and releasing the air conditioner, either the top angle is secured to the window frame by screws, or a small L-shaped bracket is fastened to the side jamb above the window to prevent win-

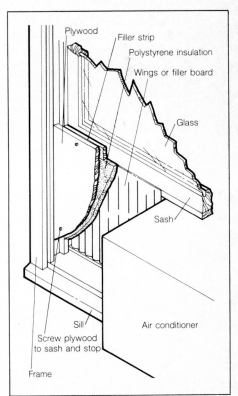

When installing a window air conditioning unit, place foam insulation between the window and the plywood filler.

Labels in diagram: Plywood, Filler strip, Polystyrene insulation, Wings or filler board, Glass, Sash, Screw plywood to sash and stop, Sill, Air conditioner, Frame

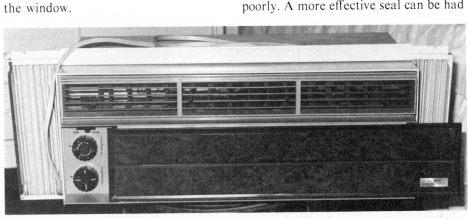

A small "Coldspot" (Sears, Roebuck) room air conditioner has a thermostat that controls both the compressor and the fan automatically, or it can be operated manually. It will cool one or two small rooms, or it can augment a central air conditioner in a hard-to-cool area, such as an upstairs bedroom.

dow movement. Place foam insulation in the space between the windows to prevent warm air and insects from entering the house.

Large Units On larger air conditioners a support bracket will be provided for mounting to the house; this relieves the load on the window frame. Position the bracket so the air conditioner tilts slightly downward to the outside of the house. This lets the condensate drain outside and not into the house. Because there usually is no provision for attaching a tube to the cabinet to drain the condensate to a specific location, the water will drip to the ground below.

Special Units for Special Windows For other styles of windows, such as horizontally sliding types, you will need an air conditioner specially designed for these windows. They are slimmer and higher than the standard unit and have a retractable frame that can be pulled upward out of the cabinet and fitted to the top of the window frame. When the frame is filled with a filler board or transparent plastic sheet, the frame seals off the area and provides support for the air conditioner. If a filler is not provided, one can be made from ⅛ inch hardboard.

Permanent Installations Many air conditioners are too heavy to remove each winter and replace each summer, so they are left in place all year. Unfortunately, the filler boards and wings provide little insulation against the cold. However, with very little effort they can be sealed for a more energy-efficient installation.

We suggest placing ½-inch polystyrene insulation over the filler board or plastic wings on the outside, then placing another board of exterior plywood over the insulation. Cut the second board large enough to fit over the window sash, so the board can be fastened in place. It may be necessary to add a strip between the plywood and the window sash, depending on the thickness of the insulation and wings. Caulk all joints. Seal the space between the window glass.

Once you have your air conditioner cabinet installed in the window, and all the gaps are filled or sealed, place the chassis in the cabinet. Attach the filter and front grill. The unit is now ready for use.

INSTALLING AN AIR CONDITIONER IN THE WALL

If you are planning to use one of the larger units to cool several rooms, it may be advantageous to mount the air conditioner in a wall rather than in a window. The unit may be located in a better cooling position than is provided by a window; also, the view from the window will not be obstructed. Placing the air conditioner in a wall is more or less a permanent installation, but the air conditioner chassis can be removed from the cabinet for cleaning and maintenance.

A disadvantage of an in-wall installation is noise. The wall acts as a sounding board and amplifies every sound and vibration created by the air conditioner. To minimize noise, some sort of cushioning must be placed around the unit to isolate it as much as possible. This could be polystyrene insulation, strips cut from a carpet or heavy duty foam insulation.

Preparing the Opening Installing a room air conditioner in a wall involves the same sort of carpentry required to make a window. Try to position the opening between studs so that the minimum number of studs will be cut. If possible, locate the opening on a side of the house that will be shaded during the hottest part of the day. When the position has been determined, locate the nearest stud and mark the size of the opening that will be required. The opening will be the size of the air conditioner cabinet plus the thickness of the insulating material that will be placed around the cabinet.

Cut away the inside wall material for the opening and on either side, back to the nearest stud. Cut away material 5½ inches above the opening to allow for a header of 2x6s placed on edge. Cut away the surface material all the way to the floor below the opening. If your room is finished with paneling, remove whole panels and cut them to fit after the hole for the air conditioner has been finished. Carefully remove the baseboard and shoe molding from the cutout area by sawing alongside the outer studs.

On the inside of the wall, remove insulation. Drill four small holes, one in each corner of the opening through to the outside wall. This will indicate where the outside wall should be cut. Make this dimension the same as the inside dimension. Use these holes as

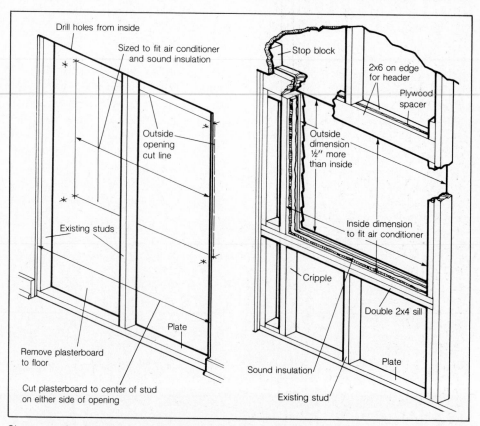

Shown are the procedures for cutting and framing an opening when installing a room air conditioner in a wall. A sill and header are placed between existing studs. Supports and spacers are added to frame the opening to fit the size of the air conditioner.

guides, so that the wall covering and its interior sheathing can be cut with a saw from the outside. On the inside, cut the studs; remember to allow for the size of the header and the sill.

Framing the Opening Cut two lengths of 2x4 to reach from the plate up to the height of the 2x4 already cut. Nail them to the flanking studs. Next, install the sill of two 2x4s, keeping the inner edge flush with the edges of the studding. Nail the sill to the 2x4 supports and toenail into the flanking studs. At the top of the opening, nail blocks to the outside studs so the bottom of the block is the same height as the bottom of the upper part of the cut center stud.

Prepare a header, two 2x6s separated by a piece of ½ inch plywood, spiked together. This sandwich arrangement gives a 3½ inch thickness, the same as

a stud. In some older homes, studs may be wider, possibly a full four inches, in which case a wider spacer will be needed. Toenail the header into the outside studs and into the end of the supporting blocks.

Finally, cut two lengths of 2x4 to fit between the header and sill and nail them to the outside studs. Add one or two 2x4 supports as needed to fit on either side of the cabinet, using spacer blocks to maintain their position. Toenail these supports to the header and sill.

Placing the Cabinet Fit the cabinet in the opening and mount the exterior support brackets so the cabinet tilts slightly downward to the outside. Seal with self-sticking foam tape and caulk the edges of the cabinet. Replace exterior siding. Replace and repair the inside

wall. Place a decorative molding around the cabinet on the inside. You may wish to frame the outside of the opening to match nearby windows. Install the chassis in the cabinet and replace the front cover; the installation is now ready.

Wiring needs If a new wiring circuit is needed, install it at the same time the inside wall is cut out. The outlet should be within reach of the electrical cord provided with the air conditioner. Do not use extension cords.

Protection from the elements Covers made of plastic or canvas may be purchased to protect the outside of the air conditioner from winter weather, or you can make a box of exterior plywood. The cover will also prevent cold outside air from filtering through the air conditioner into the house during winter months.

ADDING A CENTRAL AIR CONDITIONING SYSTEM

Central air conditioning usually can be added to any house. If your house has steam or hot-water heat, a network of supply and return ducts will need to be installed. A blower will be needed to force the cooled air to various parts of the house along with a separate thermostat. There are air conditioning units that can be installed in the attic with short runs of ducting running down through the ceiling, which is the ideal location for air conditioning to enter the room. This installation can be somewhat less expensive than locating the unit in the basement and running the ductwork under the floor and up through the walls.

A single-unit system has all its components placed in a single package and is so noisy that most people prefer to locate the unit outside the house. It then must have a short length of high and low pressure tubing running through the side of the house, connecting to the house duct system. If this system were located in the attic it would not be very efficient because of the usual high temperatures in the attic.

The system usually installed is called a "split system". The noisy condensing unit is located outside the house; the cooling coil (A-coil) is inside the furnace plenum or blower unit. Older installations required professional refrigeration

specialists to charge the system with freon after all the components were installed. New units now come precharged. Simplified connections and packaged instructions make it possible for the homeowner to install central air.

This kit includes precharged refrigerant lines, condensing unit and cooling coil.

Figuring Out the Size The first step is to get a survey booklet from the dealer, manufacturer, or the Association of Home Appliance Manufacturers. In the booklet you should list all the factors that affect the BTU gain or loss in your

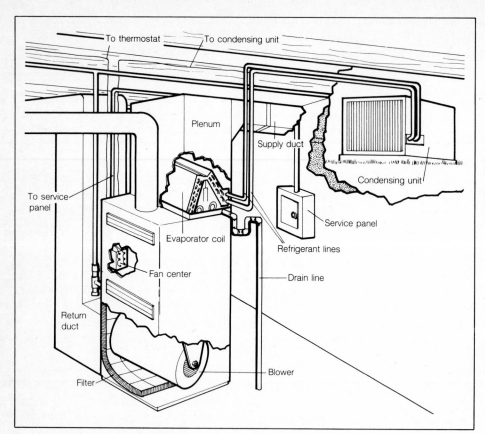

A split system, installed in a typical warm air heating system, has the condensing unit located outdoors and the refrigerant pumped to the evaporator coil through insulated lines. A blower fan circulates air from the house through the evaporator, where it is cooled and dehumidified and returned to the living area.

A level platform of poured concrete or cement blocks set in a bed of sand will support the condensing unit. It is important that it be level. Locate it so it has a free flow of air. Connect refrigerant lines and electrical lines securely.

home: the size of the house, the square footage of windows, insulation, number of occupants, appliances used, relationship to the sun and other factors. These figures determine the size of the unit needed. When this has been calculated, you can shop for the unit that gives the best Energy Efficiency Ratio (EER) for the purchase price. The EER guides you as to which air conditioner is the most efficient. Most stores will either have a chart with this information on it, or it will be shown on the air conditioner itself. If you cannot determine a unit's EER, look at the specification plate on the machine, and find its wattage. You can arrive at the unit's EER by dividing the wattage into the BTU rating (BTU ÷ Wattage = EER). The higher the number, the more energy efficient the machine. An EER around 10 is considered good, and a number of machines are found with this ratio; some manufacturers claim ratios as high as 14.

Choosing the Best Location Choose where you will place the compressor. It should be located as close to the furnace as possible, but where the noise will be least annoying to yourself and your neighbors. Check your local building codes to learn if there are any restrictions on where the compressor may be placed in relation to property lines. The compressor will need an unrestricted flow of cooling air; it should be located so the hot discharge air does not blow directly into the house or into shrubbery and flowers.

Making Tubing Estimates Precharged lines come in 5-foot increments, in lengths from 10 to 45 feet. Measure the size of the furnace plenum to be sure the evaporator A-coil will fit in it and have sufficient clearance. There should be 3 to 4 inches above the furnace heat exchanger and it should not project into the lowest duct above it more than 2 inches, or air flow into the duct may be blocked. If your plenum is not the right size, it may be possible to obtain one that is matched to your furnace and will accept the air-conditioner evaporator.

Your kit will be delivered with complete installation instructions; instructions will vary somewhat from manufacturer to manufacturer, but here are the basic steps. Always follow the instructions that come with your kit.

Sequence of Operations

Building a Base The first step is to prepare the base for the compressor, so the unit can be placed when it is delivered. The size of the base will be determined by the dimensions of the compressor. It may be a poured concrete slab 4 inches thick over 4 inches of gravel, or you can use 4-inch-thick concrete blocks set in a 4-inch layer of sand. Build 2x4 forms: place 6x6 #10 wire mesh on rocks as reinforcing. Pour the concrete, screed, and let cure. If the manufacturer requires bolts to fasten the unit to the slab, put them in place before the concrete sets. Be sure the base is level. Some compressors are equipped with leveling screws on each corner in case the base should settle and the compressor needs relevelling.

Placing the Compressor Position it on the base with the refrigerant couplings accessible to the wall. Drill access holes in the wall above the sill plate for the refrigerant lines using an electric drill with a hole saw attachment or an expansion bit in a brace.

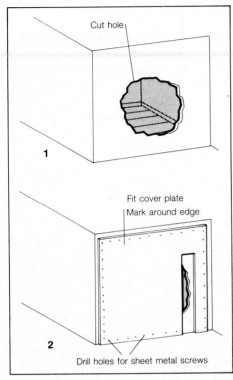

A hammer and screwdriver are used to start the hole in the side of the plenum. Tin snips are used to enlarge the hole so you can see inside and then determine the proper location for larger opening.

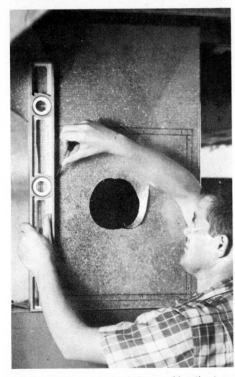

Drill holes for sheetmetal screws. Use the template provided to mark the hole and screw locations.

Preparing the Plenum Prepare the plenum for the evaporator (A-coil). Turn off the electrical power and wear a shirt with long sleeves and gloves to protect against sharp metal edges. The instructions that come with the system should show various types of furnaces and how the plenum should be prepared. Select the one closest to your design. Some will have a template to attach to the side of the plenum showing you exactly where to cut. In general, you will need to make a hole 8 to 10 inches in diameter, with the center located about 12 inches above the top of the furnace. Start the hole by driving a screwdriver or other sharp tool through the sheet metal with a hammer. Once the initial hole is made, enlarge it with tin snips. Use the furnished metal cover as a template to draw the cover's outline. With an electric drill make pilot holes for the sheet metal screws to hold the cover in position. Remove the cover and draw another mark ½ inch inside the outline

Reach through the opening and install the telescoping support rods. They fit

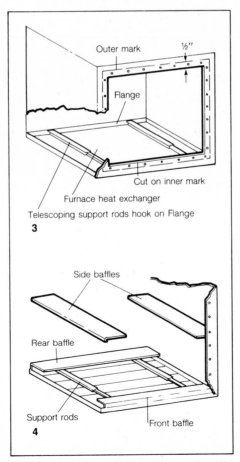

After the hole is cut in the plenum, place the telescoping support rods in position. Use caulk to seal baffle joints and the area where the cooling coil will rest.

above the furnace heat exchanger and are held in place by the flanges to which the plenum is attached. Make sure the distance between the rods is less than the width of the evaporator coil they will hold.

Cut four sheet metal baffles, provided in the kit, so that they will fill the space around the evaporator coil to form an opening the size of the coil air intake. Both side baffles will be the same size. The baffle at the front will be smaller than the one at the rear, so the fittings on the coil can protrude through the plenum. To assure air-tightness, apply caulk to the joint between the baffles and the plenum sides, and on top of the baffles about ½ inch from the inner edges of the baffles. Lower the coil in position over the baffles and press it into the caulk. Attach the plenum door with sheet metal screws.

Connecting the Lines The precharged lines are coiled for shipment;

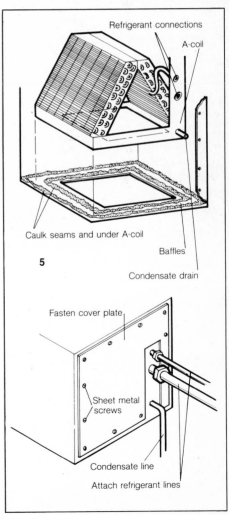

Cut the baffle and fasten it to the rods. Cooling coil is placed in the plenum so the drain projects outside.

straighten the coils carefully. Use a tubing bender to make bends as needed for installation. A bending device may be included in the kit. If not, borrow or rent one.

A spring bender provided in the kit permits bending precharged lines to produce neat bends without kinks. Insulation is placed on line after the connection is complete.

The lines are attached and tightened with a wrench. This tightening procedure punctures the seal and opens the line. Once all the lines are tightened, check for leaks with an application of liquid soap. If you find a leak, tighten the connection another ⅛ turn and recheck for leaks. If the leak persists, notify your dealer.

Attaching a Drain Line The drain line may discharge into a floor drain or sink drain that is nearby. It is possible for suction to occur through the drain line into the plenum, in which case the condensate will not drain. Attach a trap to prevent this from happening. Use a ready-made trap if available, or make one as shown.

Electrical needs The condenser requires an electrical circuit of 240 volts and, depending on the unit installed, either a 30- or 40-amp circuit breaker is required in the electrical supply lines. A safety switch must be mounted outside near the unit if one is not built in. In many areas this work must be done

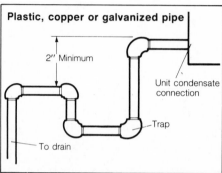

Plastic, copper or galvanized pipe

2″ Minimum

Unit condensate connection

Trap

To drain

The drain trap and pipe or flexible tubing will drain condensate from the cooling coil. Run the line to the floor drain or sump.

by a licensed electrician, so check with your local building authorities to be sure. Illegal wiring can result in a fine for the homeowner; your homeowners' insurance also could be voided.

Thermostat hookups To control the new cooling equipment, the heating thermostat must be converted to a heating-cooling unit. Some thermostats have all the controls built in, others simply require the addition of a special base to the existing unit. It may be necessary to buy a new thermostat; this may be the time to change to one with clock-controlled automatic setback. Most newer furnaces have the transformer and all necessary controls for later addition of air conditioning, so all the low voltage connections for connecting the thermostat are included. An older furnace may have only a heating transformer and it will have to be replaced with one that can accommodate the air conditioner. The connections required to hook up this new transformer or "fan center" vary considerably on different models of furnaces; you may need professional assistance to hook it up.

Some thermostats used for furnaces can be modified to control air conditioning by adding a cooling base. Here, the cooling base is hooked up and thermostat cover is ready to attach.

SETTING UP AN INSTALLATION IN THE ATTIC If you cannot connect central air conditioning to your heating system, you can still cool your house by installing a system in the attic. The condensing unit is located outdoors as previously described. The blower unit and evaporator coil go in the attic. A system of ducts is run to each room that will be cooled. A return plenum extends to a central location in the house and is fitted with a filter and removable grille. A drain line for condensate is run outdoors through the wall under the eaves. In a two-story house, ducts can be run through closets or in a boxed section in corners of a room. If it is a large house, consider using two separate condensing units located outdoors, with one blower unit in the attic to cool the upstairs, and another in the basement to cool the lower level. Ducts and blowers installed in the attic must be insulated from the summer heat.

Flexible insulated ducts are available, but if the runs are long, using rigid metal ducts wrapped with duct insulation will be cheaper. Ducts should be wrapped with the vapor barrier of the insulation facing out.

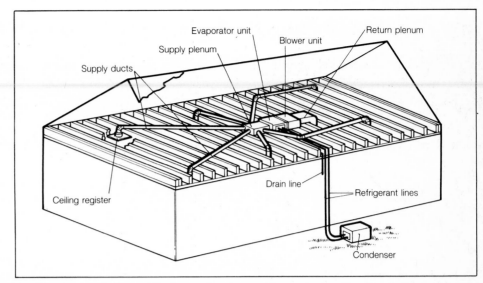

Installing a central air conditioning system in an attic means locating the cooling coil and a blower unit in a central location and running a series of supply ducts to the rooms to be cooled. The condensing unit is located outdoors. Refrigerant lines are run to the evaporator and a drain line is run outdoors through the attic wall.

MAINTAINING AND CHECKING ROOM AIR CONDITIONERS Air conditioners, both room and central, are generally trouble free. Most often, little attention is paid to them until something goes wrong. Routine maintenance can, however, keep the unit running at top efficiency and prevent problems from occuring. Before doing any cleaning or maintenance always unplug the unit or turn off the electrical power. Clean room air conditioner condenser coils (outside) and evaporator coils (inside) once a month during the cooling season. Also clean or replace the filter. Fins on the coils are thin and delicate and should be cleaned with a soft brush and a vacuum cleaner. Keep the fins on the coils straight. If they are bent, they can be straightened with your fingers, pliers or with special fin combs (straighteners). Drain tubes are another potential source of trouble. They can become clogged, causing condensate to overflow. Clean the tube with a straightened clothes hanger or soft wire.

It is a good practice to remove the chassis from the cabinet at the end of the cooling season and to clean it thoroughly. The fan usually will be encrusted with hard-to-remove dirt and may require the use of a wire brush or putty knife to remove. Do not bend the blades; use especial care if the fan is made of plastic. Bent blades and nicks can cause the fan to be out of balance and noisy.

CHECKUPS AND MAINTENANCE FOR CENTRAL AIR CONDITIONERS Cleaning procedures similar to those for room units should be followed for central air conditioners. Filters should be changed regularly. Coils, grilles and the condenser cooling fan should be cleaned annually. It may be necessary to remove a grille to gain access to the fan and coils. (Turn the electric power off before removing the grille.) Keep weeds, grass and any refuse from blocking the grill; debris will reduce airflow in and around the unit. Inside the house, clean the evaporator coil, unless it is located in a sealed plenum—in which case leave it alone. Underneath the coils will be a tray to carry off the condensate through a drain and tube. Clean the drain and flush the tray to remove accumulated dirt. Pour some house bleach through the drain to kill any fungus that may be in the tube.

When an air conditioner fails to operate properly, the cause is often due to lack of, or faulty, maintenance. In many cases, the problem is a minor one you can correct yourself. The trouble-shooting chart can be used as a guide to maintenance and repair of your air conditioner. Carry out continuity checks of thermostats, and of control or selector switches, with a continuity tester. There usually is a wiring diagram attached somewhere in the unit and it will show you which terminals to check at the various settings of the selector switch. Beyond this, if you are not trained for electrical trouble shooting, you should call in an air conditioning repairman. Capacitors can be dangerous if not shorted out prior to testing. Checking refrigerant, especially on central units, requires expensive servicing equipment. But if you have done some troubleshooting before the repairman arrives, and have ruled out some of the possible causes of the problems, you can save him time and yourself some money.

Testing Thermostat Continuity The room should be hot enough for the thermostat contacts to close. Unplug the air conditioner or turn off the power at the switch box. Label and unplug the wires to the thermostat. Attach the clip of a continuity tester to one terminal of the

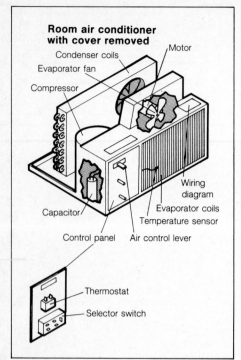

Room air conditioner with cover removed

Motor
Condenser coils
Evaporator fan
Compressor
Wiring diagram
Capacitor
Evaporator coils
Temperature sensor
Control panel   Air control lever
Thermostat
Selector switch

TROUBLESHOOTING ROOM AIR CONDITIONERS

Trouble	Possible Cause	Solution
A/C will not start	No power	Plug in cord; check fuse or circuit breaker
	Broken wire in power cord	Replace cord and plug
	Defective selector switch	Check continuity of switch; replace if necessary
Fan does not run, but compressor does run	Defective selector switch	Check continuity of switch; replace
	Defective capacitor	Check for continuity; replace
	Defective fan motor	Check continuity of fan motor; replace
Fan runs, compressor does not try to start (no hum)	Compressor overload protector contacts are open	Check for continuity; replace
	Defective thermostat	Check for continuity or place jumper across terminals with thermostat set at coldest position. If compressor runs, replace thermostat.
	Defective selector switch	Check continuity; replace
	Compressor defective	Call service technician
A/C cycles on and off or runs continuously	Thermostat set improperly	Reset thermostat
	Defective capacitor	Check for continuity; replace if necessary
	Defective thermostat	Check for continuity; replace
	Defective capacitor	Check for continuity; replace
	Restriction in liquid line	Check for excessive icing along line, indicating stoppage; call technician
	Loss of refrigerant	If apparently operating, but not cooling, charge may be low; call service technician.
Frost on evaporator coil	Outside temperature too low	Do not operate A/C when outside temperature is below 60° F.
	Dirty filter	Clean or replace filter
	Coil fins bent or dirty	Clean and vacuum coils; straighten bent fins
	Defective thermostat	Check for continuity; replace
Water leaks into room	A/C at wrong angle	Adjust unit; cabinet tilt should be ½ bubble on carpenter's level
	Condensate drain hole plugged	Clean drain
A/C excessively noisy	Exterior parts (grilles, panels) loose	Tighten screws, bolts
	Thermostat sensor bulb touching coil; refrigerant tubes chafing	Reposition sensor bulb; bend tubes slightly

thermostat. With the thermostat set at its warmest setting, touch the other terminal of the thermostat. The bulb in the tester should not light. Set the thermostat to its coldest setting; the tester bulb should light when the other terminal is touched. If the thermostat fails either test, it is bad and should be replaced with an *exact* duplicate.

Testing the Continuity of a Capacitor You will need an ohmmeter to conduct this test.

Caution: A capacitor stores a high electrical charge and causes a severe shock if not discharged before testing. Unplug the air conditioner. Discharge the capacitor by shorting out the two terminals with a screwdriver that has an insulated handle.

Procedures. Set the ohmmeter to a Rx100 or higher reading. Attach the leads of the ohmmeter to the terminals of the capacitor. The meter needle should jump toward the low-resistance end of the scale and return slowly. If there is no movement, the capacitor has an open circuit. If the needle remains at the low resistance end of the scale and stays there, the capacitor is short circuited. In either case it should be replaced with an *exact* duplicate.

Testing Continuity of a Selector Switch Unplug the unit and remove the control panel from the front of the unit. Label the wires to the selector switch and remove them. Locate the wiring

diagram, which will be somewhere on the unit. Attach one terminal of the continuity tester to the common terminal and then check each of the other terminals using the wiring diagram as a guide. Check each of the combinations with the tester. If it lights when it should not, or fails to light when it should, replace the selector switch.

Heat Pumps

Air-to-air heat pumps are not very efficient as cooling devices, but are used in milder climates. An air-to-water heat pump, where ground water at 55 or 60 degrees F. is utilized to provide cooling, is quite efficient. There are two common methods for utilizing ground water: the first involves drilling two wells close to-

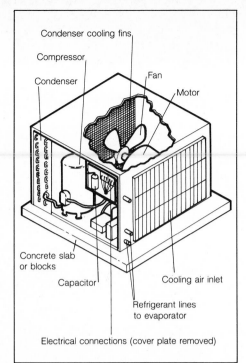

Condenser cooling fins

Compressor

Condenser

Fan

Motor

Concrete slab
or blocks

Capacitor

Cooling air inlet

Refrigerant lines
to evaporator

Electrical connections (cover plate removed)

TROUBLESHOOTING CENTRAL AIR CONDITIONERS

Trouble	Possible Cause	Solution
A/C does not run	No power	Check fuses or circuit breakers in service panel
	Thermostat set to "HEAT"	Reset to "COOL"
A/C cycles on and off or runs continuously	Thermostat set wrong or improperly located	Reset thermostat; relocate to where it best reflects average house temperature
	Compressor overheating	If compressor cooling motor operates, check if fan is tight on motor shaft. If motor not operating, replace motor
A/C runs, cools ineffectively	Dirty filter	Replace filter
	Grilles clogged, fan or blower dirty	Clean grilles, fan or blower
	Belt on blower loose	Tighten belt
Frost on evaporator coil	Outside temperature too low	Do not operate A/C when outside temperature is below 60° F.
	Dirty filter	Clean or replace filter
	Coil fins bent or dirty	Clean and vacuum coils; straighten bent fins
Blower motor overheats	Drive belt too tight	Adjust belt tension
	Motor needs lubrication	Oil motor; motors with sealed bearings must be replaced or bearings replaced.
Water leaks into room or furnace	Condensate drain hole plugged or box and pump clogged	Clean drain, box or pump
A/C excessively noisy	Exterior parts (grilles, panels) loose	Tighten screws, bolts; tape panels if necessary.
	Blower motor loose on mount	Tighten mounting bolts
	Drive pulleys misaligned	Realign pulleys
	Drive belt tension incorrect	Adjust belt tension

gether and pumping cool water from one, then discharging the heated water back down the other. The heated water is cooled by the earth. It then filters its way back to the other well, where it is pumped up to the condenser of the heat pump. The second method uses well water for cooling and discharges the heated water to a manmade lake. A third method might be to utilize the water from a natural or manmade lake to provide cooling, then to discharge the heated water back to the lake. Either a manmade or a natural lake will gradually warm up over the summer, so the

Heat pumps come in all sizes. Here two 5-ton units cool and heat a 160-year-old home that contains 10 fireplaces and 5 chimneys. The two units are housed in a brick enclosure at the rear of the house.

efficiency of the heat pump will drop. Additionally, a natural lake will have debris and algae (as will a manmade lake) that must be filtered out before the water is pumped through the condenser of a heat pump.

Devices and Products

Another device sometimes used for cool-

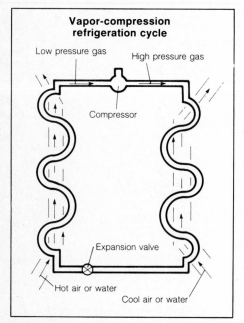

Vapor-compression refrigeration cycle

Low pressure gas   High pressure gas

Compressor

Expansion valve

Hot air or water

Cool air or water

A vapor compression refrigeration cycle—used in home and auto air conditioners, ice makers and freezers—is also found in heat pumps.

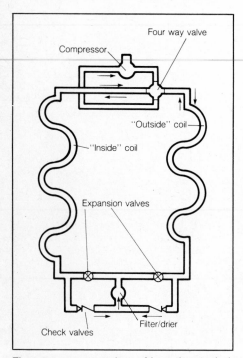

Compressor   Four way valve

"Outside" coil

"Inside" coil

Expansion valves

Check valves   Filter/drier

The vapor compression refrigeration cycle is used in a year-round heat pump. A four-way valve reroutes the refrigerant gas to provide heat or cooling inside the house. Check valves in the piping prevent the gas from backing up into the wrong pipe.

ing (and occasionally heating) is an "earth tunnel." This generally is large plastic pipe buried in the ground below the frost line. Air is pulled through the pipe, which can be several hundred feet long and is cooled naturally by the earth. Sometimes a heat exchanger (radiator) is located in the opening at the end of the earth tunnel, through which cool water from a well or lake is pumped to provide additional cooling in the summer.

One system, from the Servel Refrigeration Company, now on the market will provide air conditioning using solar energy. In the Servel system, heat is applied within a completely enclosed system, causing a phase change in the refrigerant. The resulting gas, which passes through coils and absorbs heat from outside the coils, recondenses into a liquid and starts the cycle all over again. There are no moving parts and very little can wear out. Typically a gas flame is used as the heat source but, in theory, any heat source will work. Application of solar heat is a logical phase of development.

Conventional air conditioning systems waste a lot of energy by continuing to cool the house even when the outside air has fallen below the inside house temperature. Use of attic fans and ventilation by opening windows, as was mentioned earlier, will help solve this problem. But it is possible to set up a fan system

that will automatically shut down the air conditioning and circulate cool outside air when temperatures reach an optimum level. Such a system is marketed by Sears, Roebuck and Company, but any competent heating and cooling contractor can work out this type of arrangement for you.

Fans and vents can offer dramatic heat reduction when properly planned and placed. For an in-depth discussion, see the next chapter, "Ventilation".

"ALTERNATE ENERGY" COOLING METHODS

Alternative methods of cooling have not progressed as rapidly as have alternative heating systems. Perhaps it is partially because the sun is a natural heater.

Design and Natural Cooling Aids

The design of the house can dramatically reduce the cooling load. Earth-sheltered homes, which are easier to heat in winter, are also easier to cool in summer. Better insulation, and use of masonry and shading devices, are other means of keeping the heat out. Properly designed solar greenhouse systems can assist cooling; as this air is heated, it rises and pulls cooler outside air in and a natural flow of ventilating air results. When the sun sets, other methods must be used to move the air; attic-mounted exhaust fans are the most common means.

At the other end of the scale of heat pumps is this unit that resembles an ordinary window air conditioner. It will fit neatly in a window or can be installed in outside wall, so that cost of ductwork and separate venting systems is eliminated.

3
Ventilation For Comfort And Savings

Most homeowners are well aware that insulation is needed to conserve energy. However, few are aware that when you add insulation, you also must provide adequate ventilation. In fact, ventilation becomes increasingly important as energy costs rise. During summer months, hot air can be removed from attics so the heat there is not transmitted to the living space below. Whole-house attic fans can pull warm air from living areas. Ventilation cools us by moving the air around our bodies; as a result, air that is saturated with moisture is replaced with drier air.

During winter months, proper ventilation can move the heated air that is stratified near the ceiling to the lower levels of the room. Ventilation also prevents moisture buildup in insulation in attics, crawl spaces and sometimes even in walls. Insulation that is saturated with moisture will freeze; frozen insulation soon loses its ability to insulate.

ATTIC VENTILATION
Locations and Requirements
The primary ventilation source in any home is the attic. By dealing with this area you can cut both heating and cooling bills all year long.

Summer Heat and Ventilation Summertime sun can cause the air inside an unvented attic to reach temperatures of 150 degrees or more. When this happens, the insulation becomes a "heat sink," gradually absorbing the heat until it obeys the laws of thermodynamics and moves to the cooler living space below. Often this area is being cooled by an air conditioner. An attic temperature of 135-140 degrees may force the air conditioner to run con-

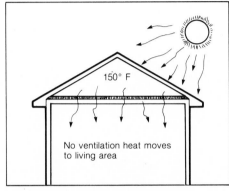

During the summer, heat buildup in an unvented attic can reach temperatures of 150°F. The insulation in the attic absorbs heat, which would be transferred to the living space below (arrows show heat flow).

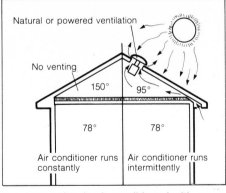

In a house that is air conditioned with an unvented attic, the air conditioner must work constantly to maintain 78° F. With ventilation and an attic temperature of 95 degrees, the air conditioner will run intermittently (the arrows show the air flow).

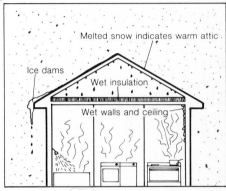

During winter, an unvented attic collects moisture that moves through the ceiling. The moisture condenses as it cools, saturates the insulation, and may leak into the walls and ceilings.

If the attic is ventilated, the moisture that escapes from the living space below is vented to the atmosphere. A properly installed vapor barrier minimizes this problem (the lower arrows indicate moisture flow; the upper arrows show air flow).

stantly to maintain 78 degrees in the living room. The insulation is now no longer protecting you from the heat as it was intended, and you have a problem that is nearly as bad as having no insulation at all.

The solution to the problem is ventilation that allows the heated air to escape from the attic before it can build

up. Aided by a roof fan or ventilation that reduces the temperature in the attic to 95 degrees, the air conditioner will run intermittently and its operation costs will be from 10 to 30 percent less.

Winter Condensation In winter, ventilation can remove humidity that moves through the ceiling as water vapor. In the cold attic, the vapor condenses in the insulation, which in turn loses most of its effectiveness. In extreme cases, icicles can hang from the rafters or the underside of a roof.

When the weather warms or the roof is heated to above freezing, the moisture melts and soaks the insulation, possibly even leaking through the ceiling to cause stains or damage. The moisture also may leak into the walls, soaking the insulation there, causing damage that may not show for years. Finally it will reveal itself in the form of rot or other structural damage. Vapor barriers in the ceiling are designed to prevent this process, but many houses do not have ceiling vapor barriers and, in any event, we cannot be assured they will be 100 percent effective.

What this means is that ventilation is just as important in the winter as it is in the summer, but for different reasons.

Means of Ventilating an Attic Natural ventilation relies upon fixed nonpowered devices called ventilators, or "vents." These are located in openings in the attic (or any other space to be ventilated) to take advantage of the natural flow of air. Cool air enters the lower vents, is warmed, then rises and moves out the upper vents to expel both heat and humidity.

Powered ventilation relies on turbine ventilators, electrically-powered exhaust fans mounted on the roof or in the attic, or on ceiling-mounted whole-house exhaust fans. Powered fans require sufficient vent area in the roof to allow the air to be pulled, or exhausted, from the attic.

Ventilation is accomplished either naturally or with power, and the method you employ will depend on several factors: (1) Is your home an existing structure or is it under construction? (2) What is the architecture of the home? (3) What are the climatic conditions where you live? (4) What are your local utility costs?

How Much Is Needed?

To determine how much ventilation is needed for your home, either natural or powered, you need to know the square footage of the attic. Then you can relate that figure to the "free area" needed for the ventilation system you want to install.

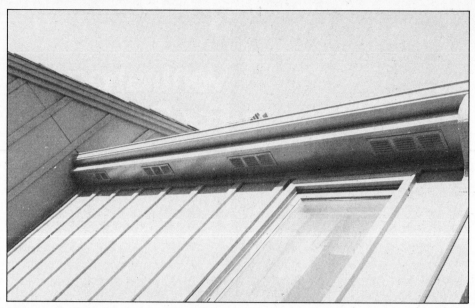

New energy-efficient homes provide soffit vents like these, or continuous vents. For maximum effectiveness, vents should be evenly spaced on all sides of the attic.

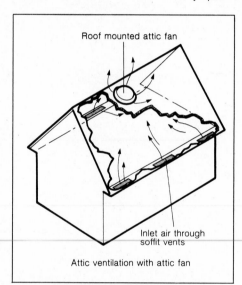

Powered ventilation accelerates the transfer of air through the lower vents and exhausts it by means of an electric fan. There must be sufficient inlet vents in the soffit for the fan to be effective (arrows show air flow).

Finding the "Free Area" Free area is the area of vent that is not restricted by louvers or screen wire. It is usually related to the square footage of the attic and is expressed as a ratio of vent area to attic area: 1/150 would mean one square foot of free vent area for each 150 square feet of attic floor. The amount of free area is determined by the existence (or lack of) a vapor barrier, location of the vents, and the climatic conditions since where the home is located determines its need for condensation and heat control.

The rules recommended by the HUD (FHA) Minimum Property Standards

(MPS) are those most often followed. Today, these figures are inadequate, but they are quoted in older publications. Homes built only ten years ago often have too little ventilation by today's standards. Below are the present generally accepted rules for determining the net free area for attics.

Attics Without Vapor Barriers One square foot of free area is needed for each 150 square feet of attic space (1/150). If half of the vents are in the eaves and half are located at least three feet above the attic floor, you can reduce the required free area to 1/300.

Attics with 6 Mil. Polyethelene Such an attic calls for a free area of 1/300. This assumes that half of the vents are in the eaves and half are located at least three feet above the attic floor.

Vaulted (Cathedral) Ceiling You will need a continuous air space above the insulation, with continuous eave vents and continuous ridge vents, or you must have individual vents in the eave as well as near the ridge for each rafter space.

Climate In areas having a high heat gain during the summer, attics with or without a vapor barrier should have a total free area of 1/150; half of the vents should be in the eaves and half near the ridge of the roof.

Vent Size When figuring the size of the vents to be used, remember that the figures above are for the net free area, which refers to the area of the vent that

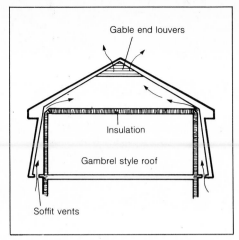

In a house with natural ventilation, the cooler air enters the vents located in the lower part of the roof or the soffits. It is warmed and rises, escaping through upper vents, carrying away moisture and helping to cool the attic (arrows show air flow).

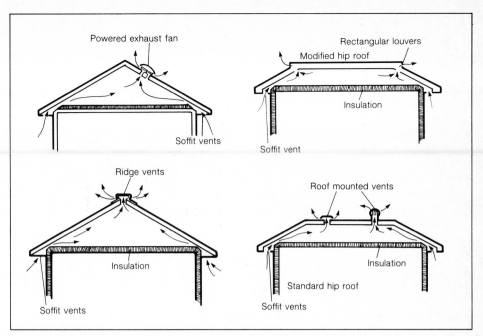

Different styles of roofs require different combinations of vent installations. (Arrows indicate air flow.) Vents in the gable end should face prevailing winds.

is not obstructed by louvers or screen wire. Use the factors in the chart below to determine the required gross area of vent openings. For example, if you are using gable end vents with louvers and ⅛-inch mesh screening, you will need 2¼ square feet of vent before you have 1 square foot of net-free area.

CONVERSION TABLE

Protective material	Net:Gross
Hardware cloth (¼ in. mesh)	1:1
Screening (⅛ in. mesh)	1:1¼
Insect screen (1/16 in. mesh)	1:2
Louvers and hardware cloth	1:2
Louvers and screening	1:2¼
Louvers and insect screen	1:3

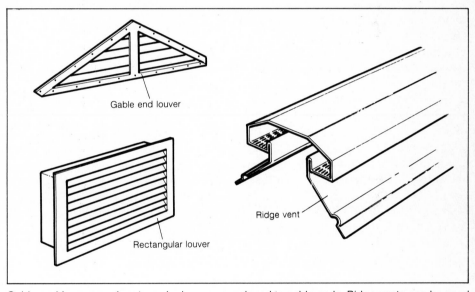

Gable end louvers and rectangular louvers are placed in gable ends. Ridge vents can be used on the ridge of any roof, regardless of roof slope. Roof vents are used on flat roofs, and are easy to install in any pitched roof. They also are used as outlets for exhaust fans. Under-eave or soffit vents have either louvers or screens to prevent entry of insects.

With a Powered Fan Powered roof or gable attic fans are rated in cubic feet per minute (cfm) at either "static air pressure" or "free air delivery." To determine the minimum cfm your home requires, multiply the attic floor area by 0.7. Add 15 percent for a dark roof. If the fan is rated in "free air delivery," discount the cfm by 25 percent to arrive at the correct figure. For example: if your attic measures 1,500 square feet, you require a fan delivering 1,050 cfm (or 1,208 cfm, under a dark roof). For best results, you must also provide adequate under-eave venting; one square foot of vent area for every 150 square feet of attic floor space is recommended (1/150).

NATURAL VENTILATION
Types of Vents
There are four kinds of vents generally used for natural ventilation: ridge, roof, under eaves and gable end. All can be used with nearly any type of architecture and, in most cases, can be painted to become quite unobtrusive.

Ridge-plus-soffit System Considered to be the best system for natural attic ventilation is the ridge vent in combination with the under-eave or soffit vent. Ridge vents provide a continuous opening along the ridge line of a pitched roof. A full flow of air can pass through

the ridge vents, but they are designed so rain and snow cannot pass through into the attic. Screens prevent the entry of insects. To complete the system, vents are fitted into the roof overhang on both sides of the roof. Installation of ridge vents used to be considered a job only carried out when a house was constructed. However, ridge vents designed for do-it-yourself installation in an existing house now are available at hardware and home center stores.

Roof and Gable-end System Roof vents and gable-end vents are often easier to install than ridge vents, particularly in older homes. Gable-end vents can be almost any shape in order to blend with the architecture. Louvers keep out the rain and snow, and screens on the inside keep out birds and insects.

Safety in the Attic and on the Roof

Many of the projects described in this chapter involve work in the attic or on the roof. Use caution when you work in the attic. Space is often limited, and it is easy to trip and fall through the ceiling. Wear a hard hat for protection against roofing nails. Provide adequate lighting, and use catwalks if your attic is not floored. If the work involves being close to insulation, or moving some around, wear goggles and a mask.

It is even more important to observe caution when you work on the roof. Use a sturdy ladder, wear rubber- or composition-soled shoes. If the roof pitch is too steep to walk on safely, use a ladder made to clamp over the ridge of the roof, or get a ridge-straddling attachment. If the roof is slate or has a pitch of over 45 degrees, call in a professional to do the work.

INSTALLING A VENT IN THE GABLE END

A gable-end installation is easy and does not require working on the roof. Begin by purchasing the correct size vent for your attic's square footage. One gable end must face the prevailing winds. Otherwise, you will not achieve the exchange of air that you need.

Tools You will need a drill or brace to bore a starter hole for a keyhole saw, a hand saw to finish the cut, plus a hammer and nails.

Step 1: Opening the Gable End On the inside of the attic, outline the area where the vent will be placed. Try to mount it as high as possible directly under the center of the gable. Cut the siding and studs from the outlined area; then cut an additional 1½ inches from the studs above and below the outlined area.

Step 2: Framing the Opening Cut and nail two 2x4 headers to fit horizontally between the studs on either side of the opening. Cut two cripple studs to fit between the headers to frame the vent. Nail the cripples to the headers.

If you are fitting a triangular vent in the top of the gable end, cut a single header to fit across the top of the cut studs, and flush against the end rafters. Nail 2x4s to fit under the double end joists from the header to the ridge.

Step 3: Inserting the Vent From the outside of the house, lay a bead of caulk around the edge of the opening. Insert the louver unit, and nail or screw it through the siding and into the 2x4 framing. Most louver units are equipped with insect screening. If yours is not, staple screen to the louver framing on the inside of the attic, or fasten the screen in place with screen moulding to keep the screen in position and to provide a neat, finished edge.

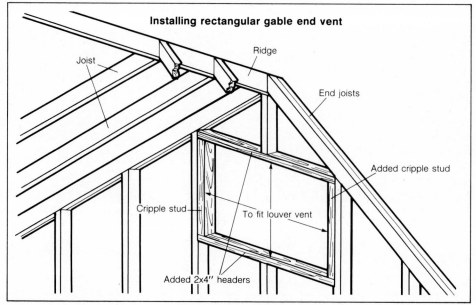

To install a rectangular gable end vent: cut the central stud; add 2x4 headers, reinforced with cripple studs on either side; fit the vent in the opening from outside the attic.

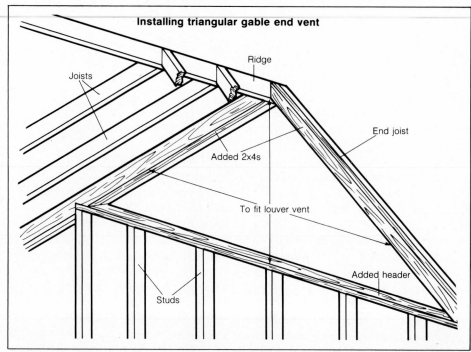

A triangular-shaped gable end vent is usually larger than a rectangular vent and will require that more studs be cut. A header is placed across the cut studs; reinforcing 2x4s are added to the end joists. The vent installs from outside the attic.

VARIATION: INSTALLING A TURBINE VENTILATOR

Turbine ventilators are an effective method of ventilating the attic using the natural "power" of moving air. The better-quality turbines use a ball bearing and are permanently lubricated. Made of galvanized steel, turbines will operate for years without maintenance. The slightest breeze causes them to operate. Even on a perfectly still day, when temperatures are high, you can see a turbine moving as the air passes through it from the attic. When used in combination with soffit vents, turbines are the second most effective natural ventilation system you can install.

Sizes Turbine ventilators come in several sizes; 12- and 14-inch diameter sizes are the most common. To determine the size and number of ventilators needed, you once again need to know the square footage of your attic. A 12-inch turbine ventilates 600 square feet. A 14-inch turbine ventilates 700 square feet. For most effective use, also install soffit vents at the rate of 1 square foot of free area for each 150 square feet of attic floor space.

Types Turbines are made with internal and external bracing. If your area is subject to high winds, the externally-braced type is recommended. In those areas of the country were it is feasible to reduce attic ventilation in the winter,

Turbine vent is a popular retrofit for natural ventilation. The rotor head uses ball bearings, so even the slightest breeze causes the turbine to draw air from the attic. The vents are effective even when there is no breeze.

turbine bases are available with an aneroid-operated automatic damper that closes at 50 degrees F. and opens fully at 90 degrees F. A useful accessory for those areas subject to hurricanes is a hurricane cap. The turbine is removed from the base and the cap put on to cover the hole.

Balancing the Turbine In the winter, a turbine can become out of balance due to snow or ice accumulation. Plastic covers are made to cover the turbines in the winter; some people place plastic garbage bags over them. This solves the out-of-balance problem, but the bags are unsightly and if turbines are the only ventilators used, the covers prevent winter ventilation of the attic. A better so-

lution is to screw a large hook on the end of a wooden stick of the right length and to hang the stick on one of the internal braces of the turbine. The turbine will not be able to turn, but natural ventilation through it can continue.

Step 1: Mounting the Turbine A turbine installation does not follow the plan of the roof. Instead, by means of an adjustable base, the turbine stands parallel to the ground, even though the roof is pitched. The base is mounted over a hole cut in the attic roof. The hole should lie between rafters and be located down from the peak. At least 8 inches of the turbine should stand above the ridge line.

Position and cut a hole in the roof equal in diameter to the hole in the turbine base. First, drill a starter hole; then use a sabre saw or keyhole saw. Remove nails holding the shingles above and beside the hole, and out as far as the flashing of the base will reach. Leave the shingles in place. Install the base and turbine.

Step 2: Flashing and Sealing Apply roofing cement under the flashing of the base and around the edge of the hole. Slide the flashing under the shingles above the hole and over the shingles below the hole. Nail the flashing to the sheathing at four-inch intervals along the top and sides. Apply roofing cement to the nail heads and to any cut shingles. Do not seal the bottom of the flashing.

INSTALLING SOFFIT VENTS

To complete the ventilation system, install undereave or soffit vents. These are sold as trim vents, either 4 or 8 inches wide and 16 inches long, or as continuous vents, 4 and 8 feet long.

Calculate the vent area and the number of vents required. Plan for vents at regular intervals in all eaves. Locate the vents midway between the edge of the eave and the outside wall, and halfway between the lookouts that support the soffit.

Outline the screened area of the vent on the soffit, and cut out the outlined area with a sabre saw. Place the vent in or over the hole and screw it to the soffit through the holes in the vent flange.

Continuous Vents Continuous vents

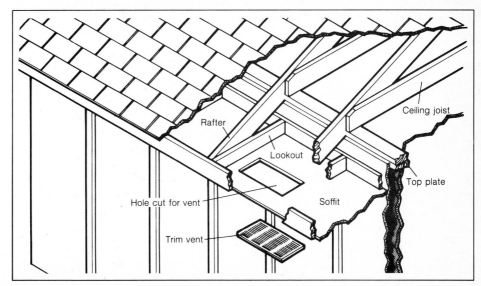

Small soffit vents, called trim vents, are easily installed under the eaves and are necessary for natural and powered attic ventilation. A hole is cut with a sabre saw and the vent is nailed or screwed over the opening.

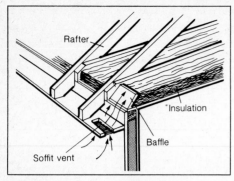

Baffles between ceiling joists prevent insulation from blocking the effectiveness of the vents (arrows indicate air movement).

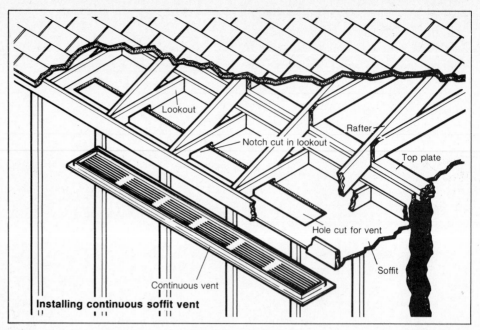

Installing continuous soffit vent

Continuous vents allow more air flow linear foot of eave. These install similar to trim vents.

are installed in a similar manner. Cut out the outlined area between the lookouts. Use a hand saw to cut the soffit where it attaches to the lookout. Some vents are recessed into the soffit, which means you may have to cut a hole in the soffit and notch the lookouts or the rafter ends. As an alternative, place 1x2 cleats on either side of the cutout, and nail the vents to the cleats.

Baffles One problem encountered with vents placed under the eaves is that insulation from the attic can cover the vents and reduce their effectiveness. Baffles can prevent this, but they can reduce the thickness of the insulation at the point between the edge of the roof and the top of the wall. Careful workmanship is important here. Install baffles or ducts that are high enough so that a full thickness of insulation does not restrict air flow through the vent.

POWERED VENTILATION

Although natural ventilation usually is adequate if gable vents are no more than 50 feet apart, power ventilation used in conjunction with the natural air flow can save energy in an air-conditioned house.

Types of Fans

A fan designed for attic ventilation is usually installed inside one of the upper vents. Some are installed in the roof itself if gable vents are not practical. Some attic fans come with vents for installation in the gable.

Energy-saving Benefits A power ventilation system not only reduces the load on an air conditioner; it also enables you to use a unit with a small capacity. The smaller the air conditioner, the less energy it requires. If you install a new high-efficiency central air conditioning unit, install a power ventilation system

This fan is designed for gable end use. With built-in framing, it can be used behind existing louvers, or with a ventilator kit installed in the gable without existing louvers.

at the same time to have even greater savings. For instance, power attic ventilation can eliminate the need to run an air conditioner at night. On those nights when the ambient temperature is at a comfortable level, open the attic door and the windows on the cooler side

of the house. The fan then pulls air through the house and expells it out the attic vents. The fan must be large enough for the job, and it should be either a variable speed or at least a two-speed type. Then, if you plan to ventilate just the attic, use a much slower speed.

When installing a fan here or in any part of the house, read all the instructions and use the correct wire size. It is better to use a wire gauge larger than called for than to use one that is too small. Undersize wire will overheat in use and be a potential source for fires inside a wall, ceiling or the attic. All switches and receptacles must also have a UL label. A few dollars more for wiring and fixtures will pay off in the long run. Appliances will work properly and more efficiently because they receive the full voltage, and the risk of fire in your home will be greatly reduced.

This fan can be mounted on flat or slanted roofs. The dome lifts when the fan turns on and lowers when the fan cuts off.

INSTALLING A ROOF-MOUNTED ATTIC FAN

A roof-mounte attic fan should be positioned as near as possible to the center of the attic; place it on the back side of the roof so it is not seen from the front of the house.

Step 1: Placement Assemble the fan and carry it to the roof near one of the gables. Using a straightedge or a piece of wood as a guide, set the fan assembly so the top of the fan is level with the roof ridge. Measure this distance from the ridge to the center of the fan.

In the attic, locate the central part and measure down from the roof peak (on the back side of the roof) to a point that corresponds to the desired location of the fan. Locate this point halfway between the rafters. Drive a nail up through the roof at this point so it can be located from the top of the roof.

Step 2: Cutting the Opening On the outside of the roof, locate the marker

A powered roof ventilator can efficiently reduce attic temperatures. This model comes with screening to protect from insects.

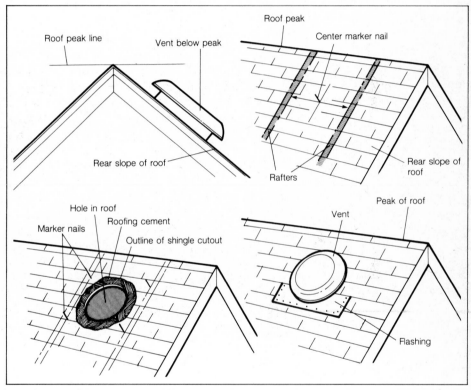

Position the fan so that it will not be seen from the front of the house. Pound a nail at the center of the proposed site. This should fall between the rafters. Marker nails will remind you of the rafters' positions. Cut the opening. Fit the flashing on the high side underneath the shingles; the flashing below the fan falls on top of the shingles.

nail. Using it as a center, draw a circle about four inches wider than the size of the hole specified in the instructions provided by the fan's manufacturer. With a utility knife, remove the shingles and underlayment down to the plywood sheathing underneath.

Using the nail as a center point, draw another circle; this should be the size specified in the instructions. Cut a hole in the sheathing or battens along this circle, using a sabre saw or a keyhole saw. The specified hole size may be larger in diameter than the distance between the rafters. If so, do not cut the rafters; saw along their inner edges.

Step 3: Installing the Housing About 6 inches above and below where the fan flashing will cover the roof, hammer 4 nails into the roof to mark the location of the rafters. Remove any shingle nails within the area above the hole that may prevent insertion of the fan flashing underneath the shingles. Apply a liberal amount of roofing cement to the exposed sheathing and the underside of the fan flashing. Slip the fan housing sheathing under the shingles above the hole. Line up the opening of the fan housing with the hole cut in the roof. Using the rafter marks as a guide, drive galvanized roofing nails through the flashing at the top and along the sides at 4- to 6-inch intervals. Using roofing cement, place a tab under any shingles that have been lifted; seal any cut edges and exposed nail heads. Do not seal the bottom edge of the flashing.

Step 4: Adding Soffit Vents Calculate the vent area you will need for your attic and install soffit vents, according to the instructions given above. For every 150 square feet of attic space, install one square foot of net free vent area.

Step 5: Connecting the Wiring Fasten the fan thermostat to a rafter so that the dial is easily accessible and the temperature-sensing element is exposed to the air. Make sure the element is not in the fan's direct air stream when the fan is operating. A 120 volt circuit using a junction box in the attic powers the fan. If there is no junction box available, run a 14-gauge copper (or equivalent) cable through a wall to a junction box in a room on the floor below.

Turn off electrical house current at the service entrance before doing any wiring. If you are unfamiliar with methods of installing electrical wiring, secure the services of a qualified electrician. All electrical connections must be in ac-

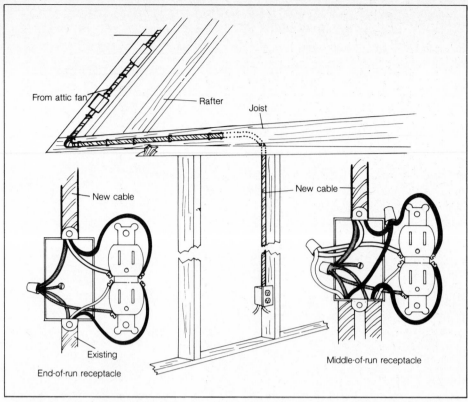

If you do not have a receptacle that can be tied into in the attic, run a line down a wall into the room below. If the receptacle is at the end of an electrical run, connect it to unused terminals. If it is middle-of-run, receptacle jumper wires must be used with wire nuts.

cordance with local codes, ordinances or the National Electrical Code.

Drill a 3/4-inch hole in the top plate above an inside wall and directly above the receptacle junction box you wish to tap. Recheck that the power is off and remove the junction box. Fish the cable through the hole in the plate to the receptacle. Clamp the cable to the receptacle, run it from the plate in the attic along the side of a ceiling joist, up the

side of a rafter, and connect it to the thermostat. Hold the cable in place with cable staples or clamps.

For End-of-run Wiring To connect the additional cable to an end-of-the-run receptacle, attach the black and white wires to the free terminals on the receptacle and the ground wire to the existing bare wire. Always use wire nuts to splice wires, not just tape.

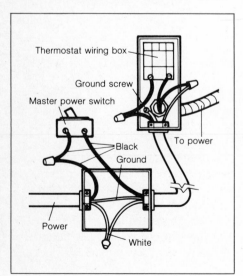

For a standard installation, connect the two leads from the thermostat wiring box to the two power leads. Attach the ground wire from the power cable to the green screw in the box. To turn the power off manually, connect a master power on-off switch, as shown here.

The fan shown is wired so it can be operated through the thermostat, or the thermostat can be bypassed and the fan operated manually through the master on-off switch.

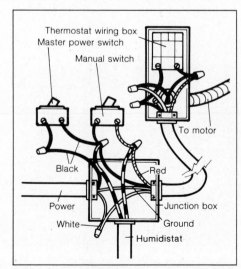

Adding a humidistat permits the option of having the fan respond to the demand of the thermostat (high temperature) or the humidistat (high humidity). It can also be turned off manually.

For Middle-of-run Wiring If connecting to a middle-of-the-run receptacle, disconnect the white and the black wire of one set of cables in the box. Cut a 4-inch jumper wire from single-counductor black wire; cut a second jumper from single-conductor white wire. Strip 1 inch from the other end. Splice together the 1-inch end of the black jumper with the black wires of the existing cable and the new cable; attach the ½-inch end to the brass receptacle terminal. Splice the white wires and the white jumper; attach the ½-inch end of the white jumper to the silver terminal screw. Finish off the splices with wire nuts; secure with electrician's tape. Connect the ground wire in a similar manner and replace the receptacle.

Connecting the Fan Thermostat After you have extended cable up to the fan, connect the two leads in the thermostat wiring box to the two power leads, matching wire color and fastening with wire nuts. Connect the ground wire from the power cable to the ground screw, usually green in color, in the box. You may wish to be able to turn the power on or off manually in order to bypass the thermostat, to turn the fan on or off manually, or wire in a humidistat. With the last installation, you have the capability of automatic operation triggered by temperature and humidity, or manual operation bypassing the thermostat and humidistat, and the convenience of easily turning the fan on or off.

Venting Requirements Before operating the fan, be sure there is sufficient soffit venting or gable-end venting to provide access for incoming air. Insufficient inlet area can overwork the fan, and it will overheat, resulting in early fan failure.

INSTALLING A FAN WITH A GABLE VENT

If you have a gable vent, a fan may be mounted inside the attic and there will be no need to cut a hole in the roof. Fans designed for gable-end mounting have flanges on the fan assembly for mounting to the rafters or studs. It may be necessary to add 2x4 supports so the fan can be mounted in the proper position. If you do not have gable-end vents, fans are available with automatic louver shutters that can be installed in the gable in a manner similar to that used for installing louvers.

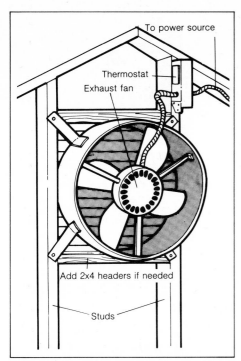

If you already have a gable end vent, installing a gable-mounted attic exhaust fan is relatively simple. Add additional headers or supports as needed.

(Diagram labels: To power source; Thermostat; Exhaust fan; Add 2x4 headers if needed; Studs)

HOW TO ADD A WHOLE-HOUSE LOUVERED FAN

Before selecting a whole-house fan, figure your ventilating requirements. The fan should be able to change the volume of air in the selected area in approximately three minutes. When you do not want to move all the air in the house, as when only one or two rooms are to be ventilated, a two-speed or variable speed fan is a good choice.

Leave sufficient clearance between a ceiling-mounted fan and the roof or other obstructions above it. Otherwise, the fan will create back pressure, which reduces the efficiency of the fan and increases its noise level. The clearance requirements vary with the size of the fan and the manufacturer, but generally are from 24 to 36 inches.

There must be enough vent area in the roof to allow the air to be exhausted outdoors with no restrictions. The vents may be gable-end louvers, roof vents or soffit vents. The net free area should be 1 square foot of opening for each 750 cubic feet per minute (cfm) of fan capacity. For example, if your fan has a capacity of 3,600 cfm, there should be 4.8 square feet of net free vent area.

If you install any other ventilating fans that exhaust into the attic, these must be ducted through to the outside. Otherwise, the whole-house fan will force air from the attic back into the living area. Kitchen range hoods and bathroom ceiling fans are included in this category.

Three accessories that should be considered for a whole-house fan are an automatic temperature control, a timer,

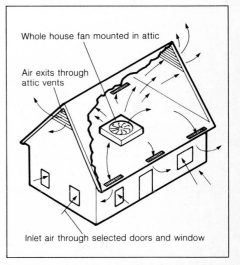

A whole-house attic fan draws air in from the windows and doors you have selected to be open, and forces the air into the attic and out the vents. It may be operated manually, by a timer, or with a thermostat.

(Diagram labels: Whole house fan mounted in attic; Air exits through attic vents; Inlet air through selected doors and window)

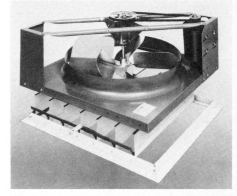

Whole house ventilators come in 24, 30 and 36 inch sizes. They are designed to provide maximum circulation with minimum rpms.

and a high-temperature switch. For safety, automatic shutters should always be used.

Step 1: Creating the Opening Mark the outline of the shutters on the ceiling

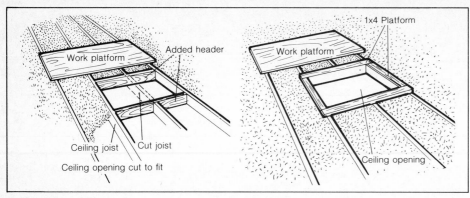

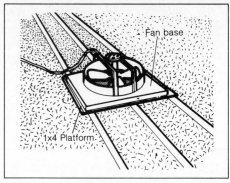

To install a whole house fan in the ceiling: mark the opening required on the ceiling; locate the nails in the attic; cut the ceiling material and joist if required; add headers between joists; build a platform of 1x4s around the opening on which to set the fan.

Place an insulating material on the platform, then set the fan in place. Nail retainer strips around the fan to keep it from moving out of position.

or wall where the fan will be located. Make the opening smaller than the actual overall size of the shutters to allow their edges to "trim" the opening. Drive four nails, one in each corner of the marked opening. Then go up to the attic and locate the nails. They should not be near electric wiring or ducts or located on joists. Either move any wiring that interferes or relocate the fan. Create the opening so there is at least 1½ inches clearance from one joist.

Step 2: Supporting the Fan A frame made of the same 2-inch stock as the joists is built to hold the fan in the ceiling. The frame can be built separately, if you have room to move it into the attic after it is built. If there is not, build the frame in place in the attic. Lay it on the joists and mark them for cutting. You can substitute a framework of headers and stub joists built into the ceiling, if you like.

Step 3: Cutting the Joists Cut the joists 1½ inch shorter on each side of the opening to allow a header to be nailed to them.

Roof trusses If your house has truss-type roof construction, do not cut the joist without first getting professional advice as to how to reinforce the truss. Some roof trusses cannot be properly reinforced, and cutting them could cause problems. One whole-house fan that does not require that the joist be cut is manufactured by Emerson Environmental Products. This type installation could be the obvious choice for a truss roof.

Step 4: Placing the Frame The frame fits between the joists and is nailed to the headers and joists with 16 penny box nails.

Step 5: Building a Frame Next, make a frame of 1 inch stock on which the fan will set. Nail this to the shutter frame.

Step 6: Inserting the Fan Place the fan in position. Some fans have a rubber or plastic cushion to minimize vibration. If yours does not, install strips of heavy weatherstripping or carpeting between a ceiling-mounted fan and the frame to minimize the transfer of vibration to the joists. Do not nail the fan solidly in place; secure it just enough to hold it in position. Wooden cleats are often used around the edges of the fan housing to keep it from moving sideways. This method will only work for ceiling-mounted fans. Those mounted in walls must be fastened more securely, but not so tightly that vibration is transferred throughout the house.

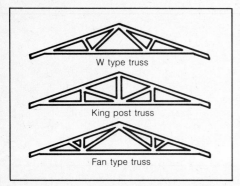

Truss roofs are constructed so that ceiling joist integrity is essential to maintaining a structurally secure roof. Cutting the joist may seriously weaken the roof structure.

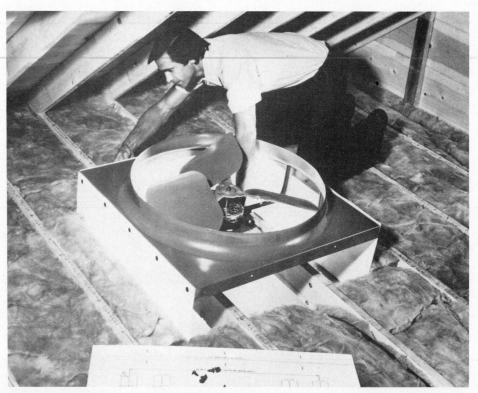

Some whole-house fans can be installed in an attic floor or wall without cutting ceiling joists or studs. Decide where the fan will be located; cut a hole in the ceiling; attach the fan to joists. Louvered shutter fits over opening.

Ceiling shutters that fit over the opening in the ceiling or wall open and close automatically with fan usage. Shutters can either be fit flush or framed if studs are not cut.

Step 7: Installing the Louvers The louver (ceiling shutter) is installed after the fan is in place. It fits in the opening from the inside of the house and is fastened with screws. The shutter is designed to open automatically with air pressure when the fan is operating. Check the louvers for smooth operation. If they do not open completely, check for possible binding or twisting of the frame.

Step 8: Connecting the Wiring Mount the fan's operating controls in a convenient location in the living area. All wiring should conform to the National Electrical Code and local regulations. If you are not familiar with the codes, get professional help. The accompanying drawing shows typical wiring for a whole-house fan. In this diagram, note the timer, which will be located near the speed control. The speed control incorporates an on-off switch. Also shown is a high-temperature limit switch, which turns off the fan in the event there is a fire in the attic or elsewhere in the house. This switch is located in the attic.

Bathroom Ventilation

Most building codes require ventilators in bathrooms without windows. Many bathrooms with windows also will have one. The window or the ventilator helps remove moisture and excess heat, as well as odors. Using an exhaust fan to remove excess moisture and heat that would otherwise build up will also lower air-conditioning energy costs. Opening a window may remove odors, but this wastes fuel in a centrally air-conditioned house.

Bathroom ventilators are available in a variety of designs to fit a wide range of bathroom sizes and budgets. Some

RECOMMENDED FAN CAPACITY
Cubic Feet Per Minute (CFM)

CFM	Laundry, Family or Recreation Room Sq. Ft.	Kitchen Sq. Ft.	Bathroom Sq. Ft.
40	50		35
50			45
60	75		55
70			65
80	100		75
90			85
100	125		95
110			105
120	150	60	
140	175	70	
160	200	80	
180	225	90	
200	250	100	
250	310	125	
300	375	150	
350	435	175	
400	500	200	
450	560	225	
500	625	250	
550	685	275	

Note: Ceiling height of 8 feet is assumed.

units combine an exhaust fan with a light; others have a combination of exhaust fan, fan-forced heater and infrared heaters. Most fans are one-speed, but models with up to five speeds are available. Fans with a squirrel cage design are quieter than bladed fans.

Ventilation engineers suggest a fan with the capacity to make from 8 to 12 complete air changes per hour. Fans are rated in cubic feet per minute (cfm).

An exhaust fan can operate from a wall switch that is independent of the

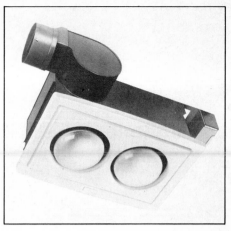

The infrared bulb heaters in combination with this fan provide instant sunshine warmth and comfort without turning up the thermostat.

light—or integral with the light. Some codes require that in windowless bathrooms a ventilator must go on automatically when the light is switched on.

Ceiling fans are installed between ceiling joists and are ducted either through the roof, the eaves or outside wall. Wall fans can be ducted directly to the outside when mounted on an outside wall, or through ducts to the roof, eaves or outside wall. The fan should discharge directly to the outside, either through a duct or a passage that is part of the fan unit. Many older installations merely exhausted into the attic. Usually a duct can be retrofitted to these fans so they exhaust to the outside.

UL Label When you purchase any ventilating fan for use in a bathroom, be sure that it is intended for use in a high-humidity area and that it has an Underwriters' Laboratory tag or label. This shows that it has passed rigorous independent testing.

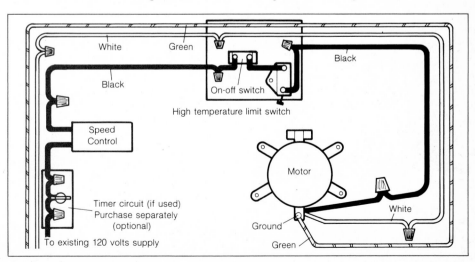

Leave the on-off switch in the high temperature box in the on position. A special switching link will automatically turn the fan off in case of an attic fire.

INSTALLING AN EXHAUST FAN ON THE WALL

Mounting the fan in a wall is similar to the technique used for a ceiling installation. The fan should be mounted as near to the ceiling as practical, since that is where the heat accumulates. Cut the opening for the switch. Above it, cut an opening for the fan. Run the cable from an existing receptacle to the switch; then run cable from the fan opening to the switch. Then complete the necessary wiring hookups.

Running the Ducts The duct can either be routed through the attic and then the roof, or out through the eaves. If the bathroom is in the lower level of a two-story house, the ducts may need to run through a wall to get to a suitable exit point. Ideally, they could be routed directly through the wall, but if there is not an outside wall, this would be impossible. Again, it may be necessary to remove part of the plasterboard and replace it. (See Chapter 1 for instructions on how to extend ducts.)

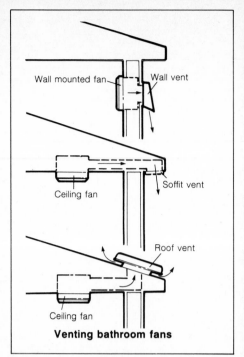

Venting bathroom fans

Bathroom ventilating fans can be mounted in the ceiling or the wall and can be exhausted through the wall, through the ceiling and out a soffit vent, or through a "jack" or vent mounted in the roof.

A bathroom exhaust fan will remove unwanted moisture and heat from the room. Choose a style that matches the room's decor.

ADDING VENTS FOR CLOTHES DRYERS IN UTILITY ROOMS

Nearly all clothes dryers are vented to the outside through a flexible plastic duct. If not vented, an excessive amount of lint will be added to the house. In summer, it will also add unwanted heat and moisture. The vent tubing should be as short as possible. It will usually be easiest to route it through an outside wall. If the laundry room is in the basement, the vent can run through the wall above the sill plate.

Since the vent will be exhausting lint, it should not be located near an air-conditioning condenser, where lint could be drawn into it. Never share a dryer vent with another vent or pipe. This can pose a fire hazard and the lint could be blown back into the house.

Heat Diverters for Electric Dryers The advantages of saving the hot humid air from a clothes dryer for use in the house in the winter time has been recognized. Now available are heat diverters (for electric dryers only) that are simply air T-valves that splice into the flexible vent pipe, and direct the air into the room or to the outside. In the winter, the valve allows the air to enter the house. In the summer, the valve is reversed and the air is routed outdoors. A fine mesh cloth filter prevents the lint from entering the room.

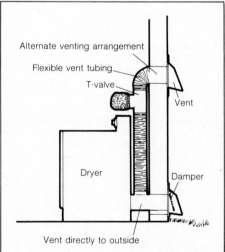

Clothes dryers exhaust great quantities of heat, an asset in the winter but unwelcome in the summer. They are typically vented to the outside directly through the wall. An alternative arrangement allows you to direct the heat either outside or inside.

Directing heat and moisture from your dryer in the winter can save many dollars each month. A T-valve that fits on the dryer vent can be closed so hot, humid air is exhausted outside in summer.

VENTILATING WALLS AND CRAWL SPACES

These two areas are often forgotten or ignored, and are inadequately ventilated.

Placing Wall Vents A house that has a good vapor barrier on the inside wall has little need for vents in the wall. However, in older homes, and in the bathroom walls in newer homes, small louver vents should be fitted above the sill and under the plate in each stud space so moisture can escape to the outside. These vents are particularly important if the siding is a vapor-resistant material in sheet form. No special venting efforts are required with wood and plywood horizontal board siding; these usually are self-venting at the overlap.

Installing Vapor Barriers Ventilation requirements for crawl spaces depend on whether the space is heated and cooled, and whether a ground cover vapor barrier has been installed. If you heat or cool your crawl space, use a vapor barrier of either 4 or 6 mil. poly-

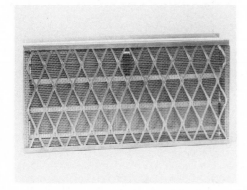

This automatic foundation vent is for crawl spaces that are neither insulated nor fitted with a vapor barrier. A thermostat causes the vent to open and close with temperature changes. At 70° F. the vents are fully open. At 40° F, they are completely closed.

ethylene film or a coated 30 lb. per square foot or heavier roofing felt or roll roofing. The edges of the vapor barrier should be lapped and taped or sealed with ballast and turned up approximately 6 inches on the walls. If the crawl space is then properly insulated along the foundation walls, no ventilation is required.

Crawl spaces that are not heated or cooled should also have a vapor barrier. The space should be ventilated with not less than two vents, located as high as

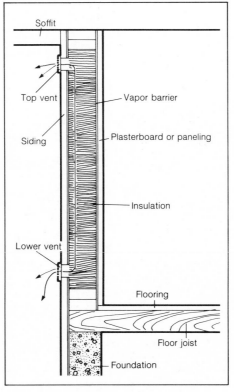

Wall vents are needed in houses that do not have vapor barriers in the walls and in all homes where the siding is vapor resistant. The arrows show air movement into the lower vent and moisture-laden air moving out the top vent.

possible and on opposite walls at a ratio of 1/1500.

In older homes, where the space is not heated or cooled, the vapor barrier can be omitted, but the ventilation should be a ratio of 1/150 and vents placed in all crawl space walls. The overlying floor should be properly insulated and have a vapor barrier (see Chapter 9).

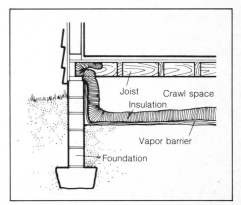

Crawl spaces that are heated and not ventilated need a vapor barrier over the ground and up the wall of the foundation. Insulation is fitted over the foundation walls and over the ground around the edge of the crawl space.

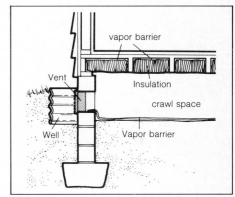

If crawl spaces are not heated, they should be ventilated. A vapor barrier is recommended on the ground, with insulation attached to the floor with the insulation facing upwards.

ADVANTAGES OF DECORATIVE CEILING FANS

We have all been in air-conditioned buildings or homes where the temperature was at the recommended level yet we perspired and felt hot. However, if the air were moving, even slightly, we would feel more comfortable. Since a ceiling fan uses relatively little power, the fan can save a relatively large amount of energy that otherwise would be used for the air conditioner.

Likewise, during the winter, there are times when we are comfortable while we

are sitting, but are uncomfortably warm when we work near the ceiling. The ceiling fan corrects this situation by circulating the warm air that has risen up near the ceiling down to where we are "living." As a result, the furnace need not run as long, thus saving energy used for heating.

Ceiling fans are available in many different designs and sizes, with variable-speed controls and even with built-in lights. The fans are quite often mounted in place of the ceiling light, so that an integral light becomes more of

a necessity than a luxury. The fans are particularly useful in homes with high ceilings.

A ceiling fan will increase the efficiency of a wood stove. A stove gives off a great deal of heat, but unless the unit is equiped with a blower, the heat will rise and become trapped in one area. A ceiling fan will move heated air away from the stove and into adjacent spaces. The fan can make the difference between a comfortably warm, wood heated home and a home that is super-heated near the stove and frigid elsewhere.

INSTALLING A DECORATIVE CEILING FAN WHEN WIRING RUNS PARALLEL TO JOISTS

Ceiling fans may be flush-mounted surface-mounted or installed with a swag kit. The ceiling junction box to which the fan is hung should be fastened securely to a ceiling joist or to a 2x4 that is supported perpendicularly between the joists. Do not mount the fan to a junction box that is supported only by plasterboard. If the fan is mounted to an existing ceiling light fixture that is operated by a wall switch, the fan is simply mounted to the junction box with the hardware provided with the fan. If it is a variable-speed fan, a new switch is installed in place of the old one. If there is no switch-controlled ceiling fixture, it may be necessary to run an electric cable from the ceiling box to a wall box.

The alternative is to use a swag kit. Here, the electrical wire is run through a chain that is routed across the ceiling and down to a wall receptacle. Operation of the fan is controlled by a switch on the fan housing.

Tools and Materials For this job you will need an 18-inch drill bit, usually called an electrician's bit, 2 fishtapes, and a saw.

Step 1: Cutting the Opening Locate the ceiling joists in the desired area of the bathroom where the fan will be placed. Mark the area to be cut out, using the template that is provided with the fan. You also can make your own, using the fan housing as a guide. Whether you cut the opening so it is centered between joists or adjacent to a joist will depend on the mounting method used with the fan. Follow the directions provided.

Step 2: Running the Cable You will have to run an additional electric cable to the fan, unless you are replacing a ceiling light with a fan-and-light combination. If the fan operates independently of the light, you will need a separate switch. If the unit is equipped with a heater, you may need a separate circuit to the main service panel.

Stringing cable through attic If you are running a cable from an attic, it is a relatively simple matter to run the cable along the side of a ceiling joist, or under the attic flooring to the point in the ceiling in the room where the fan

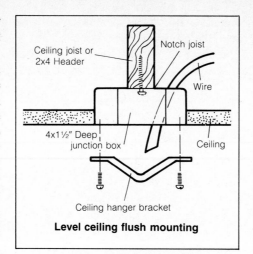

Level ceiling flush mounting

Ideally, the junction box for a ceiling fan is attached with wood screws to a ceiling joist after the ceiling surface is cut out.

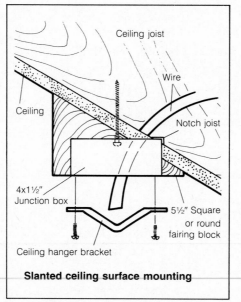

Slanted ceiling surface mounting

Because the fan should hang level, a junction box on a slanted ceiling must be mounted with blocking and a notched joist.

is to be placed. A cable also will need to run through a plate and into the wall for the switch.

Stringing cable through basement Running the cable from or through the basement involves cutting through a sill plate into the wall and up to the ceiling.

Stringing cable in 2-story houses Two-story homes present the biggest challenge, as the cable must be fished through the wall and the ceiling, whether you start from the basement or the attic. In most two-story homes you will have to fish wire through a wall and ceiling that are closed in with plaster or plasterboard.

Step 3: Cutting Openings A few inches down from the ceiling, cut a hole into the wall where the switch will be

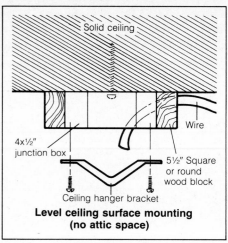

Level ceiling surface mounting (no attic space)

You may also frame around a junction box that must be mounted on the surface of the ceiling. The box attaches to the joist.

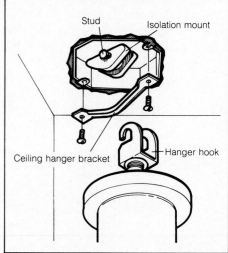

An isolation mount installed in the junction box absorbs motor vibration and limits the sound transfer to the framing.

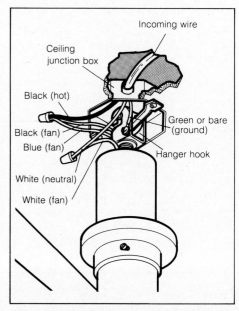

Wiring is hooked up between incoming power and the fan as shown. If your fan does not have a light, you will not have the blue wire.

A decorative ceiling fan will move heated air away from the ceiling and back down into the room, thus decreasing heating needs.

located for the ceiling fan. Drill diagonally up through the hole and through the top plate into the ceiling cavity. Cut the opening for the ceiling fan.

Step 4: Fishing the Cable Insert a fishtape up through the hole in the ceiling and a second one through the hole in the wall. It may take some work to get the two tapes to interlock so you can pull the fishtape in the wall through and out the ceiling hole. Tape the new cable to the fishtape and pull it through until it extends from both openings.

When fishing lines through wall and ceilings, keep the openings aligned if at all possible.

When Wiring Runs Perpendicular to Joists

First, cut the hole for the fan. Then make the opening for the wall switch. Cut another opening in the wall directly above the switch and immediately below the ceiling. Mark a line from this latter opening to the hole in the ceiling. Locate each of the ceiling joists on this line and cut out the plaster and notch the joists

for the cable. Carefully feed the wire above the ceiling through each notch. A length of coat hanger wire, and an assistant, will make the job easier.

At the corner of the wall and ceiling, notch the top plate and feed the wire down into the wall. Pull the wire taut and use a staple at each joist location. Patch the holes with patching plaster or plasterboard.

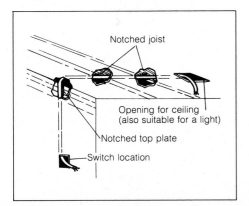

To control a ceiling fan from a wall switch, run cable from the switch to the ceiling opening. For cable perpendicular to joists, notch joists and cover cable with protective plates.

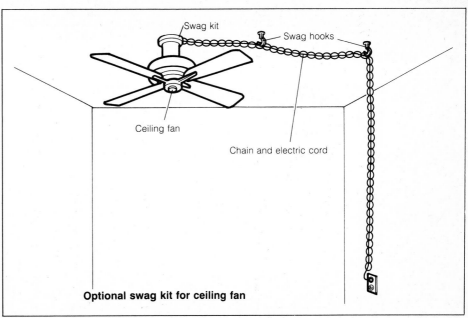

Optional swag kit for ceiling fan

If you do not wish to run power cable to the ceiling fan location, many manufacturers offer swag kits that allow you to mount the fan on the ceiling but run the cord to an outlet.

This project, designed by Frank O. Gehry of Santa Monica, California, was an American Institute of Architects 1980 honor award winner.

4

Solar Conversions: Can You Retrofit?

SORTING OUT THE TRUE STORY

Because there is so much controversy about solar energy and whether or not it is practical or economical, the first thing we will do is discuss some of the myths and misunderstandings about this form of heating a home.

First, solar heating is practical for any home, in any climate where there is sunshine at least a few months of the year. Obviously, the more the sun shines the more solar energy there will be to collect. This means that in cloudy areas, such as in the northwest, you won't collect as much solar energy as you will in the desert southwest where the sun shines almost every day. Despite the low percentage of sunny days, however, there are solar systems in Washington State and Oregon that do collect enough solar energy to make a genuine contribution to lower the energy requirements of the homes on which the systems are installed.

Cost of Installation

If you hire a qualified contractor to install a solar heating system on an average home, the cost will run from $3,000 up to $15,000. The system should have two or three panels that heat your domestic hot water the year around, and provisions should be included to shut down all the space heating collectors when they are not needed. If you are an experienced do-it-yourselfer, you can install a solar system for about half the cost of a contractor-installed setup. But there is a lot of work involved, and it will take time.

Tax incentives One big advantage is that, between federal and state tax incentives, you can figure that 40 to 50

This solar home has 324 square feet of PPG Air Collector panels and utilizes phase change salts for storage.

In an experiment sponsored by the Electric Power Research Institute, solar collectors were mounted on a south-facing roof of a solar home located in Wading River, New York.

percent of the cost of the system will be refunded to you.

HOW IT ALL WORKS
The "Greenhouse" Effect

One way to prove to yourself the viability of solar energy as a workable concept is to leave your car parked outside on a cold winter day while the sun is shining brightly. When you get into the car, and the temperature outside is hovering around zero degrees, the inside of the car will be 80 degrees F. or warmer. This same situation occurs when you sit in front of a south-facing window on a sunny winter day; a great deal of warmth comes through the glass.

The basic cause of the greenhouse effect is that sunlight passes readily through glass in the form of short wave radiation. When the light waves contact objects on the other side, the radiation changes to long wave energy. The glass is almost opaque to the long wave (heat) energy, so the heat is held inside the structure.

Preventing Heat Loss

It is true, unfortunately, that when solar energy is not constantly pouring through glass to create heat (as on cloudy days or at night), the heat rapidly dissipates through the glass, which has no insulating properties at all.

This heat loss problem can be handled in two possible ways, or by a combination of the two methods. First, the glass can be turned into an insulator by using two or three panes of glazing. This offers more resistance to the passage of heat and combats the principle that heat passes from the warm to the cold material.

The second way to solve the problem of heat loss is to store the heat that is collected during the sunny hours. Then you can move it into the area to be warmed at times when the sun is not providing heat. In some cases it is not even necessary to use mechanical means to move the heat to the cooled space. If you have a glassed-in porch that is exposed to the sun, consider the benefits of a concrete floor or a floor covered with concrete and ceramic tile. The mass of the floor will absorb heat from the sun, then return it to the room as the air in the room cools. This again is the result of the basic law that heat moves from a warm material to a cold one. In this

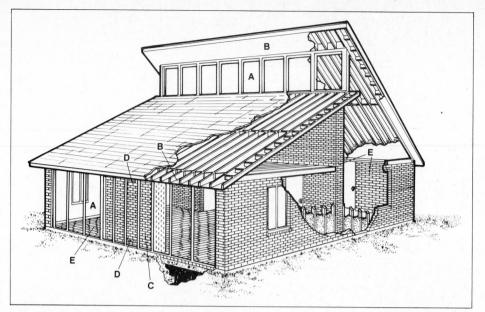

Components shown are: A, collector; B, shade devices; C and E, thermal storage; D, vents.

Passive solar heating principles are utilized in this home. The "sun space" at the far end of house is a solar collector and greenhouse. The chimney has glazing in front so it acts as Trombe wall, creating heat and storing it in the brick mass of the chimney.

case, the cool material is the air in the porch.

PASSIVE SOLAR HEATING

There are two basic forms of solar heating: active and passive.

How It Works

Passive systems are the least expensive and complicated and can be as simple as a window facing south to capture the rays of the sun that is low on the southern horizon in the winter. Concrete or adobe walls or concrete floors retain the heat so that the house itself becomes a solar collector.

Adapting a Passive System to Your House

Retrofitting a passive solar system to a house requires only large areas of south-facing glass. The windows will collect enormous amounts of heat. However, storing that heat for use at night or on cloudy days is a problem. Heat storage requires mass. This can be a concrete or ceramic floor or wall. Floors are not too much of a problem, but building a heavy concrete or masonry wall is not practical in most cases. An ordinary home is not built strongly enough to support a mass that might weigh thousands of pounds.

One storage answer has been to use black-painted steel oil drums. Other homeowners have set out containers of water; collections of empty gallon bleach bottles and the like are frequently used. However, water weighs about 8 pounds per gallon, and a volume of water large enough to store a practical amount of heat becomes heavy. Nor are steel drums or plastic bottles very attractive furnishings.

Window Arrangements The window for a passive system should be as large as practical and should have double glazing. Triple glazing is better for cold climates. The installation can consist of sealed double pane of insulating window glass with a storm sash outside the window. There is some loss of efficiency in this arrangement, in that the double and triple glazing do slightly resist the passage of the sun's rays through the glass. However, the loss of efficiency is slight compared to the amount of heat that double- or triple-paned glass holds in.

With windows as collectors it would be necessary to open and close drapes, or to open and close insulating shutters or other insulating devices in order to prevent collected solar heat from passing back through the window to the colder outside air. Ordinary window shades do an excellent job of insulating a window; use of insulating materials for the shades increases their efficiency even more.

Orientation Also, the window (or house) does not have to face absolutely south to act as a collector. A difference of up to 30 degrees east or west of true south makes only a minor difference in collecting solar energy. There are even houses that have collectors facing east and west, so they collect solar energy only half a day each. Controls, manual or automatic, shut down the east collectors when the sun no longer shines on them, while the controls start up the west-facing collectors once the sun's rays reach them.

ACTIVE SOLAR HEATING SYSTEMS

An "active" solar heating system relies on one or more flat plate collectors to absorb the heat. The plates may be located on the ground, but usually are found mounted on the roof. The plates

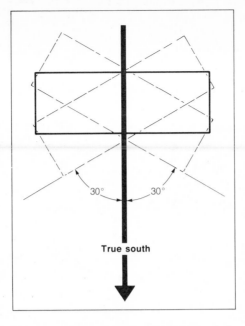

True south

face south in order to catch more of the sun's rays.

The Plate Collectors

The flat-plate collectors consist of at least one cover plate, either of glass or plastic. The sun shines through the cover plate onto the absorber, which is painted black or dark green. Tubing containing a liquid is soldered to the collector. In-

sulation is placed behind the absorber. Pumps and controllers are necessary for circulation of the air or liquid that collects the heat so that the air or liquid is directed to the heat storage area or living areas.

Materials The flat plate is built using tubes of copper, aluminum or steel. The best cover-panel material is glass—because it lasts longer and is cheaper than plastic (plastic clouds over time). However, plastic is easier to work with.

Air vs. Liquid Circulating Element Liquid (either water or, in cool regions, water/antifreeze mixture) offers an efficient heat transfer. It needs less energy and takes up less space than a system that uses air-circulation. Air systems call for more complicated storage and distribution arrangements.

Retrofitting an Active System

Active solar systems can be retrofitted to a house, and storage of heat can be handled less awkwardly than with passive systems. A liquid-cooled system can send its excess heat to a heavily insulated tank of water located in the basement or buried underground. An air-cooled solar heating system stores heat

Instead of installing solar panels on a roof, stand them along a south-facing wall of your home. With the proper trim, they will fit right in and will not look like an eyesore.

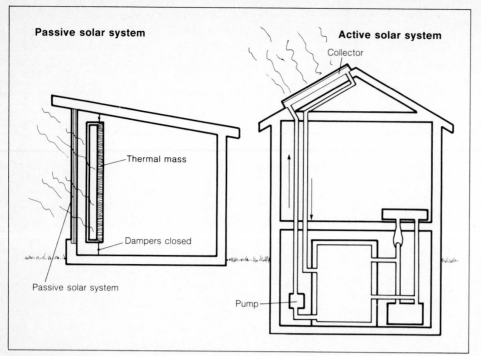

Passive solar system

Thermal mass

Dampers closed

Passive solar system

Active solar system

Collector

Pump

A liquid-cooled flat plate solar collector is an active system, in which the heat exchanger prevents contamination by the water/antifreeze mixture. A flat plate air-cooled solar collector has a blower to circulate heated air. Shown are simplified versions of systems.

in a bin of rocks, again located in a basement or underground. In the air system, dampers in the ductwork route the heated air to the inside of the home or to the storage medium.

Thermostats and Controllers

Differential thermostats are used with both air- and liquid-cooled active solar heating systems. Temperature sensors, attached to the collector outlet and storage outlet, signal the controller to start or stop the pumps. When the temperature of a collector is 3 or more degrees warmer than the house or storage, the system is activated. Pumps move the liquid from the collector to a heat exchanger that transfers the heat to the inside of the home. When temperature of the home reaches the home's thermostat setting, the heated liquid is rerouted to the storage medium.

There is a high-limit sensor on the hot water outlet of the storage tank to prevent too high a storage temperature. Pump speed is varied by some controllers to assure optimum collection of solar heat.

About 50 percent of the energy used by a hot water tank is to maintain the temperature of the water at some predetermined temperature, for example about 140 degrees F. Water this hot is required only by an automatic dishwasher; for any other use the water is tempered with cold water, thus simply wasting the energy. Water at 120 degrees in the hot water tank is perfectly satisfactory for most cases, as few people use water hotter than about 105 degrees.

LIQUID-COOLED SYSTEMS
The Plumbing System

A liquid-cooled solar heating system is a complete plumbing system, which is separate from the conventional plumbing system of the house. The system consists of the collector on the roof or other location, the piping from the collector to a heat exchanger and/or a storage tank, plus necessary pumps and controls. In modern solar systems there will be a central "brain box," which is a solid-state microprocessor. To this unit are connected thermostats in the supply and discharge lines at the collectors and at the storage tank and also in the living space.

The brain starts up the pump when solar heat warms the water in the collectors to a predetermined temperature. The heated liquid then is sent to the heat exchanger to warm the house if it is below a certain temperature. When the house is warm enough, the liquid is then rerouted to storage. In some situations, liquid will be routed both to the storage

and to the house. This will occur when there is enough heat being collected to supply both locations.

Water vs. Antifreeze Solutions

Liquid-cooled systems use either water or antifreeze/water solutions. The problem with water-cooled solar collectors is water freezes at 32° F. In areas where temperatures fall below freezing, a liquid-cooled collector must contain a water/antifreeze solution. The type of solution you use affects the type of system you must install.

Using Antifreeze Mixtures One major complication with a liquid solar heating system that uses a water/antifreeze solution is that the mixture most often used (ethylene glycol) is toxic. It must be kept isolated from potable water in the home. This is especially critical when a solar heating system heats domestic hot water. In these cases, a heat exchanger, basically a coil of tubing inside the water tank, gives up its heat to the water in the tank. A system that has a heat exchanger is called a closed system, discussed below.

Heat Exchangers You can purchase a "solar hot water tank" that has a built-in heat exchanger. Two connections on the tank route liquid from the solar collector through the exchanger coil. One other connection supplies cold water to the tank. This water is heated, then run either directly to the plumbing system of the house or to a conventional hot water tank.

If the heated water is supplied to an existing tank, the water then is raised in temperature, if necessary. For instance, on an overcast day, water from a solar system might run only 90° F. The conventional heater then increases the temperature to 115° or more.

Open and Closed Systems An "open" system is one that is under atmospheric pressure. This means that a larger pump is required to refill the system when the sun shines the following day. Not only must the pump overcome pipe and connector friction in the lines, there is considerable head pressure (the weight of the column of water the pump must raise). This is due to the force of normal air pressure on the system. An open system also is subject to air pollution and any debris that is present in the air around it.

Solar Usage NOW, Inc. offers a heat exchanger control module that hooks up to a standard tank that is purchased locally, to save freight costs.

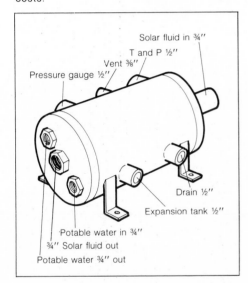

Solar fluid in ¾″
T and P ½″
Vent ⅜″
Pressure gauge ½″
Drain ½″
Expansion tank ½″
Potable water in ¾″
¾″ Solar fluid out
Potable water ¾″ out

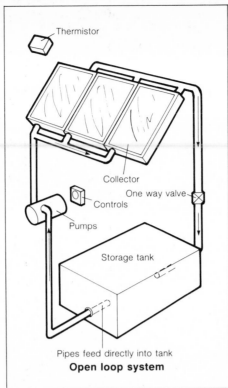

Thermistor
Collector
One way valve
Controls
Pumps
Storage tank
Pipes feed directly into tank
Open loop system

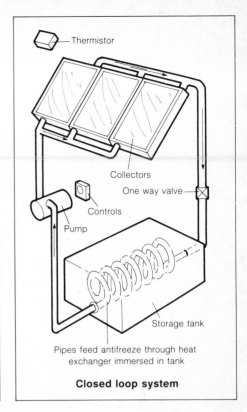

Thermistor
Collectors
One way valve
Controls
Pump
Storage tank
Pipes feed antifreeze through heat exchanger immersed in tank
Closed loop system

In a "closed" system, which is not open to the atmosphere, the pump is aided in raising the water to the upper parts of the system because there is a slight vacuum on the "down" side of the piping, this vacuum helps lift the column of water.

Types of Antifreeze Solutions
Ethylene glycol, which is the most commonly used antifreeze for liquid-cooled solar heating systems is fairly reasonable in cost, is slightly heavier than water and boils at 387° F. Pure ethylene glycol freezes at about 8½° F., but when mixed with an equal part of water the freezing point drops to −33° F.

A related chemical, propylene glycol, not as readily available and more expensive than ethylene glycol, is not toxic. It is somewhat heavier than water and has a boiling point of about 369° F. When propylene glycol is mixed with an equal weight of water, the solution has a freezing point of −26° F.

Pumps for the Two Systems A so-

lar unit using a water/antifreeze solution requires a larger pump than a unit using pure water. The specific heat of water is one BTU per pound degree F. The glycols, on the other hand, have about half that specific heat. This means that a water/antifreeze mixture must move through a solar collector faster in order to absorb the same amount of heat absorbed by pure water at a slower rate of movement. As a result, the pump or pumps used in a solar unit utilizing water/antifreeze solution must be larger than those in a system utilizing pure water.

In addition to the lessened heat transfer, the glycols have a higher viscosity (are thicker) at lower temperatures, and a thick liquid requires more pressure to pump than a thin liquid like water. The pumps for a system that uses a water/antifreeze solution must provide twice as much pressure as those for a pure water system. It is miscalculations in factors like these that cause some solar systems to work poorly (or in some cases, not at all).

Using Water Instead of an Antifreeze Solution
The danger in using water is that it freezes. One way to avoid use of an antifreeze is to have a "drain down" water system. In this arrangement, a ther-

mostat senses when outside temperatures get down around 36 degrees F. The pump automatically stops and all the water drains into a storage tank inside the house. When the outside temperature increases, and solar energy heats the inside of the collector above freezing, the thermostat tells the pump to refill the collector system. There must be an air bleed at the top of the system to allow the air to be forced out of the system as the water moves up, just as there must be a valve to allow air to enter the system. This is an "open" system, involving the disadvantages mentioned earlier.

This type of system can become complicated, but if you are willing to open and close a few valves after reading a thermometer that tells you when the water coming out of a collector is getting cold, you can save a lot of time, trouble and cost. After you have worked with a solar system for a while, you can increase the sophistication of the control system as little or as much as you want. That is one of the advantages of building and operating your own system: you learn as you work on it and operate it, and can adjust it to fit your needs.

Product Suggestion: The "Solaris System" One system that uses straight water, is straightforward in construction and has functioned successfully for many

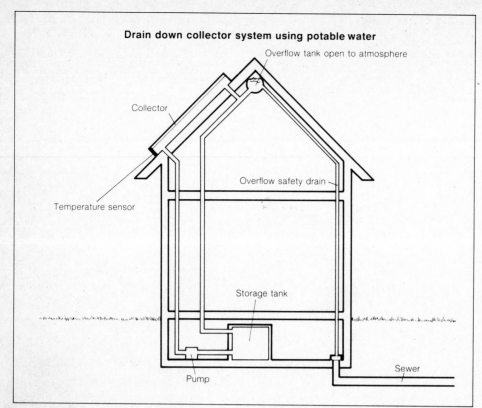

Drain down collector system using potable water

Overflow tank open to atmosphere

Collector

Temperature sensor

Overflow safety drain

Storage tank

Pump

Sewer

A drain down collector system has no valves. The amount of water allowed in the system depends on storage tank size. If the pump stops, water drains back to the storage tank. Addition of a check or stop valve would cause standing water, which could freeze.

years under actual working conditions is the system designed and patented by Harry Thomason. This "trickle" system consists of collectors that are corrugated sheets of metal painted black. Water is pumped to the upper end of the collector where it flows down by gravity to a gutter at the bottom, then down through a drain into the storage area or through a heat exchanger. Plans for this patented "Solaris" system can be purchased, and the cost includes a one-time license to use the system, building it for yourself alone.

The Solaris incorporates a drain down arrangement when the air temperature drops to near freezing. When the pump stops, the water drains back to the storage tank. Because it uses only potable water, the Solaris system requires no heat exchanger for heating domestic water, only for transferring heat for space heating.

To install a Solaris system one must purchase manufactured solar collectors from a company licensed by Thomason. This is to assure that the system will work properly. Although the system is available to skilled do-it-yourself homeowners in "kit" form, the license agreement you receive expressly states that

Mr. Thomason is not responsible if the system does not work correctly. Having seen solar systems installed by "experts" that do not work well, or at all, this qualification is understandable.

A solar contractor licensed by Mr. Thomason will install a system for somewhere between $6,000 and $10,000. A skilled homeowner with plumbing experience can reduce the cost to almost half if he does the job himself.

A Combination of Active and Passive

A "hybrid" system that combines both solar and active is the most efficient and economical setup. Thus, plenty of south-facing windows, plus liquid- or air-cooled collectors on the roof or in a south-facing wall, make the most economic sense.

SIZE REQUIREMENTS

If people tell you they have a solar system that is so good that they have to open windows in the winter time to release excess heat, they are mistaken. They have paid for a system that creates more heat than the house can handle or the system can store. This means they paid extra for oversized equipment.

Over the years, solar engineers have

worked out many tables, charts and projections, and the figures come down to a "rule of thumb." If a building is well insulated, with walls having R-20, ceilings having R-30 and basement walls having R-15 insulating factors, then a collector area equal to about 25% of the building's floor area will provide from 60 to 70% of the heating requirements. If the insulation in the house is less than these R-factors, then the collector area will have to be increased. It is less costly and complicated to increase insulation, when that is possible, than to increase collector size. Generally, you can reach attics to increase the ceiling insulation, even if you cannot open your wall space to add more insulation. Storm sash, weatherstripping in doors and windows and caulking of joints in the outside structure will increase the heat-retaining capacity of the house.

Even if a solar collector system is made big enough to provide 100% of the space and water heating needs, the problem of storage complicates the matter. On some days, the sun does not shine, and of course there is no light at night. Storage to provide more than a few days of heat is not practical because too much space would be required. This is when you need a backup system.

USING A HEAT PUMP

Heat pumps work on the second law of thermodynamics, which states that heat always flows from a warm body to a cold one. Recent design improvements have made the heat pump available today a highly efficient and very reliable one. The initial cost is comparable to a more traditional heating system.

How It Works

A heat pump works on the vapor compression-refrigeration cycle, just like air conditioners for home and automobiles. There are four parts of a heat pump system in addition to the necessary piping: an evaporator, an expansion device, a compressor and a condenser. The expansion device can be either a valve or a capillary tube that is like a long nozzle with a very small inside diameter. When the mixture of gas and liquid that is being pumped through the system leaves the small opening, it expands into gas and absorbs heat as it expands.

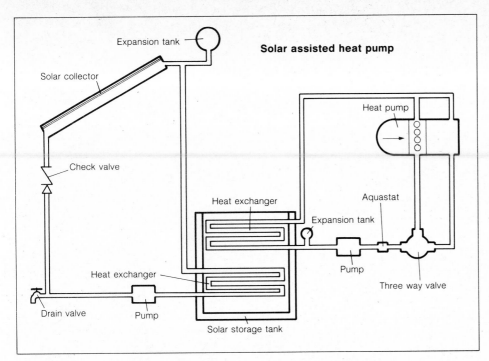

Solar assisted heat pump

The system of a heat pump (like an air conditioner) is filled with a refrigerant that usually is a fluorinated hydrocarbon with an appropriate boiling point. Probably the best known of these fluorocarbons is "Freon," the trade name owned by the DuPont Company. In operation, a heat pump or air conditioner works like this.

Liquid refrigerant boils in the evaporator, and absorbs heat from the surroundings. The evaporator is called a "finned tube direct expansion (DX) coil," and the surroundings are the mass of air blown through it by the pump's fan. This part of the system is what cools the room and you.

The refrigerant then leaves the evaporator as a mixture of liquid and vapor. This mixture now passes to the compressor, then the condenser. After the mixture has been compressed, it leaves the compressor at a higher temperature than the air or liquid surrounding the condenser. Passing through the condenser, the mixture gives up heat to the air or liquid. In the usual window or auto air conditioner, the mass around the condenser is the outside air. (It is obvious that the air around the condenser is heated by the action of the refrigerant; there is a blast of hot air from the condenser as the fan blows through it.) As the refrigerant gives up its heat, it returns to a liquid state, and the whole process of evaporation, compression and condensation begins all over again.

What has just been described is a heat pump in the cooling mode for summer. For wintertime operation, the heat pump operates as though the condenser were moved inside and the evaporator moved outside. The fan blowing air through the condenser to get rid of the heat now blows the heat into the house to provide space heating. It would not be practical to physically move the condenser and evaporator inside or outside depending on the season, so the change is accomplished by controls and piping. The condenser and evaporator now are called "inside" and "outside" coils because each changes its function from evaporator to condenser and back, depending on the season. The outside coil may actually be inside the house and in the same cabinet as the inside coil.

Types of Pumps

If the fluid around each coil is air, the unit is called an air-to-air pump. It is called a water-to-air heat pump if the outside coil is surrounded by water. While water-to-air heat pumps generally are used in large buildings, which have an ample supply of water, more and more of this type are being installed in homes, especially those where a solar-heated liquid can be used to increase the efficiency of the pump in cold weather.

Purchasing Suggestions Try to buy a heat pump that has a high EER. The higher its EER number, the more efficient it is. SEER is a more common term

listed with some appliances, and means Seasonal Energy Efficiency Ratio. The SEER usually will run about one point below the EER, because over a season the temperature and humidity will vary from the standard condition used for testing.

When and Where is a Heat Pump Feasible?

When the heating efficiency of a heat pump is determined it is called Coefficient Of Performance, or COP. Although higher COPs are theoretically possible, most manufactured heat pumps are rated at 4. Water in the range of 60 to 95° F. generally is used for water-to-air heat pumps, so their COPs remain basically constant. Unfortunately, air-to-air heat pumps work with a fluid (outside air) that varies in temperature over a wide span, possibly from 110 to 10° F. Keep in mind that the outside coil of the heat pump is pulling heat from the surrounding air to boil the refrigerant. Air contains less heat as the temperature drops, so the COP of the heat pump also drops. The chart shows how the performance of a high-quality three-ton heat pump varies with temperature changes. To relate BTU's with heat pump capacity, one ton of cooling is equal to 12,000 BTU/hr.

HEAT PUMP EFFICIENCY

Outside Temp. (°F.)	Capacity (BTU/Hr)	Kw	Coefficient of Performance
−10	7.5	2.0	1.10
0	11.0	2.3	1.40
10	14.5	2.5	1.70
17	17.0	2.6	1.92
20	18.0	2.6	2.03
30	22.0	2.8	2.30
40	27.5	3.1	2.60
47	31.0	3.3	2.75
50	33.0	3.3	2.93
60	39.0	3.5	3.27
70	46.0	3.8	3.55

Above 40° F., the chart shows that the heat pump has a fairly high COP. In most homes, very little heating is required above 40 degrees. Because the COP drops with the temperature, at 0° F. the heat pump is really not much more efficient than straight electrical resistance heating that has a COP of

The builders of this home anticipate that they will need to resort to conventional heating methods only on occasion. The passive system incorporates a south-facing wall of glass, 24 in. styrofoam insulation and a cement and rock foundation for thermal storage.

just 1. Total heating capacity drops with the temperature also, so that it usually is necessary to provide supplemental heat from a backup system. Most heat pumps are integrated with an electric resistance setup as a backup. This can be expensive.

Considering these factors, it is obvious why heat pumps are most favored in areas where the winters are mild and the summers are hot. However, in a cold climate, an extremely high electric bill one month in winter could be acceptable if the remaining months result in a reasonable heating bill because you have used the heat pump.

Solar Heating Plus Heat Pumps

Combining solar-heated water with a water-to-air heat pump can keep the COP of a heat pump high. Generally, water in storage for a space-heating solar system will be above 100° F. This can be used in a heat exchanger with a blower to heat the home, especially if passive elements are included in the design of the house. When the water drops below 100° F., it can be routed through a water-to-air heat pump that then can produce heating for the home until the water drops below 60° F., when the COP drops off. At this point a backup heating system must come on line. A standard furnace or wood burning stove would be the usual such backup. This might occur at night or during cloudy weather when there was not enough solar energy to heat the water in storage.

Cooling with a Heat Pump

When a water-to-air heat pump is used for cooling, the condenser requires a source of water to cool the unit. Very few of us have a backyard pond or running stream from which we can get water, so a cooling tower is required. Before the 1960s, almost all refrigeration used cooling towers because they are more efficient than air-cooled condensers. The towers are "closed systems" in most cases, recirculating the same water over and over. Periodic maintenance is required for cooling towers: they must be cleaned and an algae-

A Sun Saver II unit from Sun-Day Solar, Inc. fits into a window without structural changes.

cide introduced to the water to kill algae and other organisms that thrive in the hot water. Their maintenance needs and their bulk is probably the reason for the switch to air-cooled condensers, but their greater efficiencies suggest we investigate their potential. They can be disguised as backyard fountains at ground level or hidden on roofs behind dormers.

5

Building Your Own Solar Collector

A flat-plate solar collector draws on a wealth of mature technology. This device has been in use for over 50 years; in the 1920s and 30s, flat-plate solar collectors were used by the tens of thousands for heating domestic hot water.

ACTIVE SOLAR HEATING SYSTEMS
Components and Materials
A flat-plate collector consists of an absorber plate inside a box that can be made of wood or metal. Under the absorber plate is at least 2 inches of insulation. If the collector is liquid-cooled, there will be tubes attached to the absorber plate. If the collector is air-cooled, the back of the absorber plate will have passages through which air is blown.

The top of the box is covered with glazing. Glass is preferred, although some rigid plastics also are employed for glazing. Low-iron glass transmits about 90% of the solar energy that strikes it, while high-iron glass passes only about 80%. Low iron glass looks blue when viewed on edge, while high iron glass has a green look. For most applications, double glazing is used to minimize heat loss by conduction. If two sheets of high iron glass are used for glazing, then only about 64% of the solar heat is transmitted through to the absorber plate. There is a tradeoff in efficiency here; you lose less heat by conduction, but you harvest less solar heat. In cold climates, double glazing is a must; just the movement of the wind across the glazing can pull an enormous amount of heat from the glass.

Heat Storage
Rock and water are the two common materials used as heat transfer mediums for storage. One other possibility is used of Glauber's salts, sodium sulfate decahydrate. This "phase change" material heats to 90 degrees F; the salt then melts and absorbs heat. Later, when heat is taken from the salts to warm the house, it reverts to crystalline form. The problem with these salts is that the phase change sometimes stops occurring and heat no longer can be stored. The source of the problem is precipitation. To prevent it, the salts are sold in tall, thin containers. Plastic is used for the containers because the salts corrode metal.

Tubing For a Liquid System Collector
A handy homeowner can make his own liquid cooled solar collectors using any one of a variety of tubing materials. However, we recommend a recently introduced aid for the do-it-yourself solar enthusiast—the D-tube made by the Mueller Brass Company.

The D-tube is a half round copper tube with one flat side. This flat side

Phase change salts offer a highly efficient storage medium. To prevent them precipitating and changing, phase change salts are contained in small-diameter, long plastic tubes.

assures excellent heat transfer between the tube and the absorber sheet. These D-tubes are used for risers in a solar collector, and are joined at the ends to headers. Because drilling a D-shape hole is more than a little complicated, the manufacturer makes headers with the D-shape holes already punched in them. If you don't want to use these special headers, the Mueller Company also makes adapters from D-shape to round to fit the round holes any of us can drill in a round header.

You cannot buy the D-tube directly from the Mueller Company, because as a manufacturer they sell only to wholesale firms and O.E.M.'s (original equipment manufacturers). There are, however, a number of mail order outlets in the solar specialty field who will sell the tubing to you, including both the risers and headers. The same outlets also sell pumps and controllers for an active collector system.

Piping for the System

Because copper tubing is used for the heat exchanger, copper also should be used for the piping of the collector system. It does not matter what metal is used for the absorber panel because there is no liquid connection between the copper collector and the absorber panel to set up a galvanic condition.

Plastic piping should not be used for a solar system because at "stagnation," which is the condition just before the pumps start, the liquid in a collector can reach 300° F. or more. At this temperature, plastic plumbing will sag, distort and solvent-welded joints can come apart.

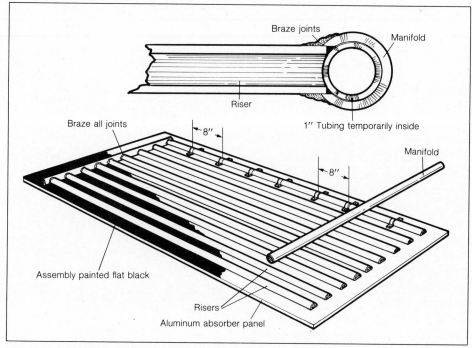

The latest development in do-it-yourself liquid cooled solar collectors utilizes D-shaped tubing, which provides greater heat transfer from the absorber plate.

PREVENTING GALVANIC ACTION AND CORROSION

Whenever possible use all the same metal in the piping for a solar collector system. When this is not possible, as when an aluminum solar collector is connected to copper or steel plumbing lines, some sort of "sacrificial anode" must be included. This arrangement is similar to the anodes used in standard hot water tanks. In that situation, a stick of magnesium is inserted in the water tank so it is destroyed by galvanic action rather than having the tank or piping corrode.

Using Screen Wire For a solar heating system you can, in the case of aluminum collector panels, use aluminum screen wire as the "anode." The wire is fitted in a section of the piping that is larger than the main lines. If the lines to and from the collector are ¾ inch, then the section of tubing or pipe in which the screen is held should be 1 inch. The section is inserted either with reducing unions, which are not always

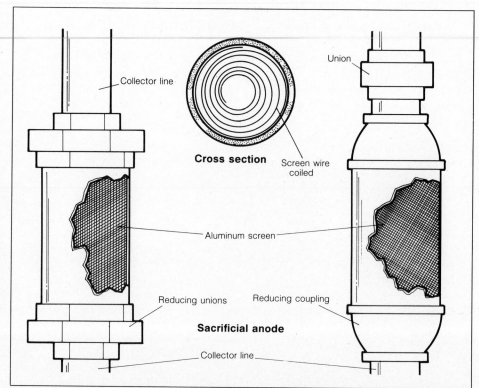

Pipe for the sacrifical anode is one size larger than the collector piping. When the anode wears out, the unions or couplings permit easy replacement.

easy to find, or with reducing couplings, unions and nipples. The aluminum screen wire is rolled into a cylinder and slipped into the larger section of pipe.

The screen wire will last for several years, which is about when the antifreeze in the system should be replaced.

Because the antifreeze is the same as used in automobile cooling systems, it must be replaced on the same sort of schedule as auto antifreeze.

Changing the Antifreeze Cooling systems in cars may combine copper or aluminum radiators with cast iron blocks and aluminum heads and the antifreeze does not cause a problem as long as it is changed on a regular basis. This is important: when a car becomes a few years old and starts to overheat and the radiator must be cleaned and the block "boiled out," it is because corrosive action has plugged up the fine tubes of the radiator and the passages in the engine block and head.

Antifreeze has additives that maintain a neutral pH factor, but these additives gradually wear out and the antifreeze turns acid. The acid attacks the metals in the system and causes corrosion. As in an automobile, antifreeze in a solar collector system will turn acid, especially when subjected to the high temperatures produced in a full sun. To assure that the antifreeze in a system has not become acid, you can test the water every 6 months to a year with a

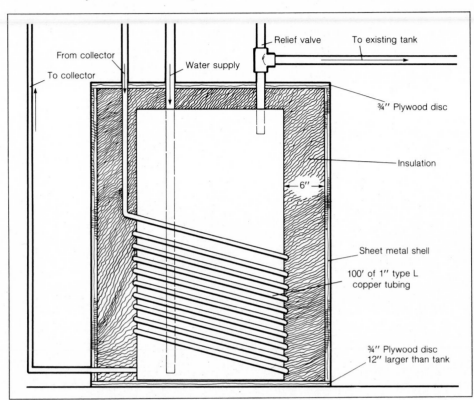

This is a standard 82-gal. water storage tank, around which copper tubing is wrapped to bring heated water from the solar collector.

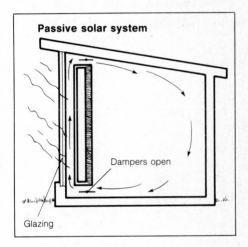

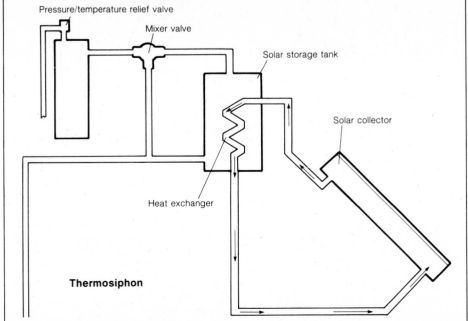

In a thermosiphon system, the storage tank must be at least 12 inches above the top of the collectors so that heated water will travel up to it, causing circulation to force cooler water back down through the collectors. This is like a passive system in that it has no pumps; otherwise, it functions like a small active system.

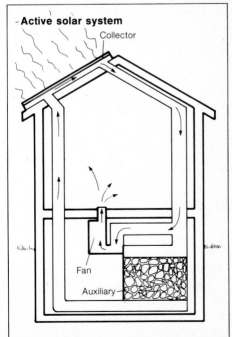

pH kit, similar to the one used for checking the water in a swimming pool. If it has turned acid, drain the system, flush it with clear water and pump in fresh water/antifreeze solution.

This may seem expensive, but there are not too many gallons in a solar system, and spending a few dollars for fresh antifreeze is a lot less expensive than replacing piping or collectors.

Guidelines for Use of Aluminum Collector Panels

Aluminum collector panels should be used only with distilled or deionized water, even when a water/antifreeze solution is required. If you live in an area with "hard" water, which means there is a high percentage of minerals in it, this water could cause problems the moment it is pumped into a system that contains more than one kind of metal. Some hard waters can be deionized, by means of a water softener. This works by exchanging sodium ions from salt with the hydrogen ions in the mineral-containing water.

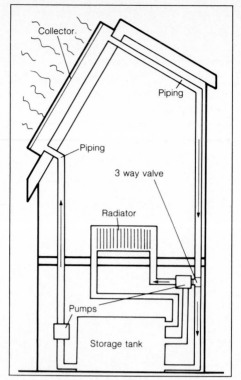

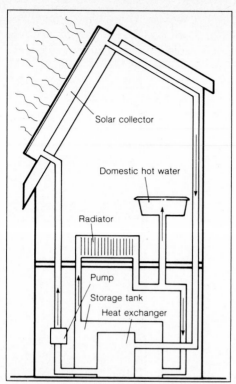

A liquid-cooled collector's heat exchanger can be a radiator or baseboard heater. Differential thermostats control a 3-way valve, for storage of heat when tank temperature drops.

For a simple heat exchanger, place a tank within a tank. Outer tank can be cast concrete, sealed inside with waterproofing. Inner heat exchanger tank can be steel or fiberglass.

BUILDING A FLAT-PLATE COLLECTOR

Remember that these "kits" of tubing for risers and headers do not usually include an absorber plate, and you will have to make the plate from aluminum, steel or copper.

Step 1: Assembling the Plate Considering efficiency and cost, aluminum is the best buy. Use aluminum sheet .025 inches thick (the exact alloy is not important). You can buy the material from aluminum products distributors who have it in rolls. It usually will come in 36-inch widths, but you can have it cut to size, or cut it yourself with tin snips. To fit the collector assembly shown, make the sheet 33x72 inches.

Brazing risers and headers The copper risers and headers should be joined with brazing rather than soldering. This kind of brazing sometimes is incorrectly called "silver soldering," although neither silver nor solder ordinarily are in the alloy.

A brazing alloy such as BCuP is inexpensive. It runs about half the cost of 50-50 soft solder and one third the cost of 95-5 tin-antimony solder. Also, the

Heat manifold and riser until solder melts when touched against hot metal. Use a propane torch or a propane/oxygen setup. Keep flame on tubing, not on stick of solder.

The finished solder joint should show a complete fillet all around, between the smaller riser and the larger manifold. Practice until you can produce a smooth solder joint.

brazing alloy requires no flux. If your supplier doesn't know what BCuP brazing alloy is, tell him you want hard solder for joining copper tubes.

Small hand-held propane torches will not provide the necessary 1300 to 1500 degree heat required for brazing, but you can use an oxy-acetylene torch or even a hand held MAPP® gas torch. Propane torches such as the Wingersheek "Turbo-Torch®" utilizing a special nozzle also can be used.

Don't be afraid to try brazing. Practice on a few scraps to see how the brazing rod melts when touched to the heated joint and flows around it, then join the risers and manifolds in the pattern shown.

Testing risers and manifold Each assembly of risers and manifold should be tested for leaks before it is attached to the absorber plate. You can use either air or water for this, although for most of us the water test is easier.

You will need a short length of auto radiator hose with about 1⅛ inch inside diameter, a reducer, a female garden hose connector and a couple of hose clamps.

Fill the riser-manifold assembly as full as possible, attaching the hose to one end of the manifold. Then plug up the other three openings. Turn on full water pressure and carefully inspect each brazed joint for leaks. Pressure in most city water mains is 50 or 60 pounds per square inch, and the relief valve in

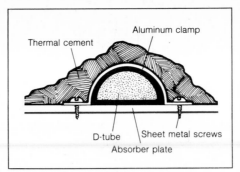

the collector system will be set to open at from 8 to 15 pounds per square inch, so a water test should be satisfactory. When air is used from a compressor, the compressor generally is set at 100 p.s.i.

Attaching risers to the plate To provide good heat transfer between the risers and the absorber sheet, it is necessary to use a heat transfer cement. The best is Thermon® and you can locate your nearest distributor by writing to the Thermon Manufacturing Company, 100 Thermon Drive, San Marcos, Texas 78666. The thermal cement also is sold by solar mail order houses. (One of the best mail order sources is "Peoples' Solar Sourcebook". You can get their catalog of over 350 pages by sending $5 to Solar Usage Now, Inc., P.O. Box 306, 420 East Tiffin Street, Bascom, OH 44809. Not only is this a catalog of all kinds of solar components, it also is a textbook on solar installations,

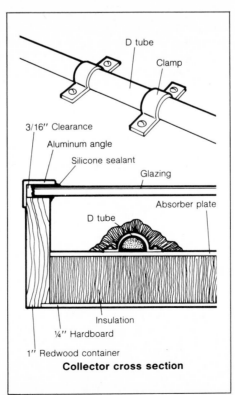

Collector cross section

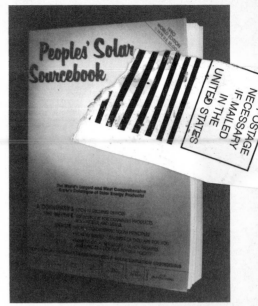

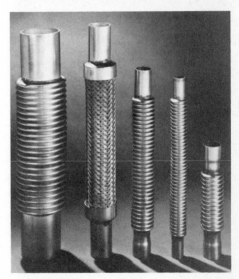

Flexible tubing comes in many sizes to connect solar panels and meet other piping needs.

with drawings showing various ways that solar collectors can be connected, and descriptions of the various controllers required for a system. If you are interested in solar, the catalog is well worth its price.)

Thermon thermal cement is available in standard nonwaterproof grade T-3 in 1- and 5-gallon cans, and in a waterproof grade T-85 in 1.10 gallon cartridges for use in a caulking gun. The waterproof grade is stronger, but costs about four times as much. It also is smoother, since the T-3 has the consistency of chunky peanut butter. Both will provide good results in a solar collector and saving money means a bit more work in application.

While the thermal cement is excellent for heat transfer, it is not a good mechanical bond. Hold the risers to the absorber plate with cleats cut from the aluminum that you trimmed off to make your absorber plate the right size. Cut each cleat about ½ inch wide and about 3 or 4 inches long. To shape the cleats, bend them around a scrap piece of the D-tube. Then space them about every 8 inches on the risers, and fasten them to the absorber plate with sheet metal screws.

Run a bead of Thermon® along the absorber plate where each riser will contact it, then attach the risers with the aluminum cleats. Apply more of the thermal cement over the D-tubes to increase heat transfer. If the T-3 formulation is used, you can figure about two-thirds gallon of the material for each collector.

Step 2: Building the Container

Redwood generally is used for the container. We suggest 1x6 lumber ripped to a width of 4½ inches and rabbeted as shown.

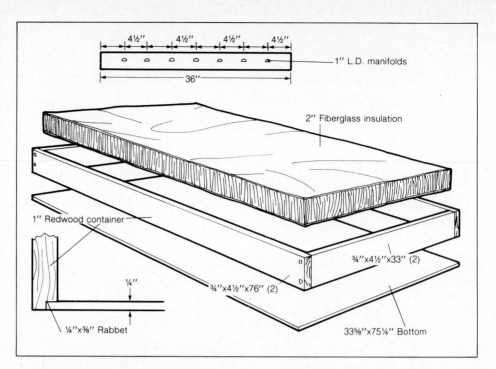

The back of the wooden container is ¼ inch tempered hardboard or sheet metal. Assemble the container as detailed, using waterproof glue and non-rusting brass or aluminum wood screws.

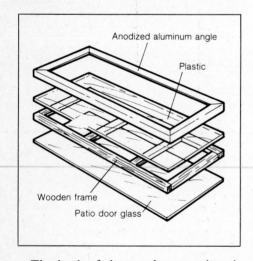

There has been some controversy lately about the possibility of wooden containers burning due to a buildup of heat from the absorber, since wood degrades under heat. We have never known of a spontaneous ignition of a wooden collector container, although we are aware of some that have been operating on houses for 5 or 6 years. Also, temperatures in an average attic can climb to 150 degrees and higher, but we have never heard of spontaneous ignition of rafters or ceiling joists in a home.

However, if you have some prior experience with sheet metal, you can make your collector containers of metal. In this case you will have to insulate the sides of the container as well as the bottom. Metal does not have the insulating properties of wood. Commercial collector containers are made of metal for the most part, but this is partly due to the lower cost of metal fabrication as opposed to wood.

If two or more collectors are to be installed, it makes sense to make one large container, and thus save on material and reduce the exposed surface of the container. Where two or three collectors are used for hot-water heating and several more for space heating, then you will want to divide the installation into two containers. Thus, in the summer when only the water-heating units are used, you will have no heat carryover to the space-heating collectors.

Step 3: Adding the Insulation

At least 2 inches of fiberglass insulation is fitted between the bottom of the container and the collector plate. You also can add insulation to the edges of the container. This is not absolutely necessary in a wooden container, but is in a metal one. High-temperature insulation is best (it can be mail-ordered from the Peoples' Solar Sourcebook if not available locally). Don't use Styrofoam® or urea formaldehyde rigid insulation. These will break down under the high temperatures created in a solar collector.

Ordinary fiberglass home insulation has a binder that gasses off at about 300 to 400 degrees and this could cause fogging of the collector cover plate during a "stagnation period." This occurs when the collector is heated, but the pump has not been turned on to remove the heat.

Step 4: Painting the Collectors

Paint the collectors after they are installed in the containers. The inside of the container and the edges of insulation also are painted at the same time. We feel that the best paint to use is Nextel Black Velvet, manufactured by 3M. However, other paints are almost as efficient and cost much less than the $60 a gallon for Nextel. For even greater savings, use a flat black paint, such as formulated for barbecues, engine exhaust systems and the like; the paint is sold in hardware stores and auto supply outlets.

Step 5: Installing the Cover Plate

If a standard tempered patio glass door is used as a cover plate (you can buy the glass replacements for patio doors in a standard 34x76 inch size), place the glass on a bed of silicone sealer that has been laid all around the top edge of the container.

Apply a bead of the sealer to the top edges of the glass and hold the glass by screwing on angles of anodized 1½ inch x 1½ inch aluminum. The bronze anodizing has a much richer appearance than plain aluminum. The anodized material can be obtained from glass and storefront remodelers. A minimum clearance of 3/16 inch should be allowed between the edges of the glass and the aluminum to allow for expansion. The flexible silicone sealer will accommodate the movement of the glass as it expands and contracts due to heating and cooling.

Creating a double plate You can make a double cover plate to produce higher water temperatures at a cost similar to the single glass cover by using "Filon Type 546" plastic. Make the frame as indicated but add two intermediate supports (you can use the strips that were ripped from the 1x6 redwood pieces) for the plastic. One sheet of plastic is screwed to each face of the frame and sealed with continuous beads of silicon sealer. In addition, use brass or aluminum wood screws spaced about 6 inches. (Try to do this assembly on a day when the humidity is low.) Apply enough sealant so that it oozes out when the screws are tightened.

Attach this cover to the edges of the redwood container, using additional sealer around the top of the container, along with brass or aluminum screws.

The flat plate collector you have built has an area of about 17 square feet. To supply most of the year-round water-heating requirements, for an average family, you will need four of these collectors.

Step 6: Positioning the Collector

In addition to plumbing and soldering experience, physical agility is needed if you decide to place the panels on the roof. Solar collector panels should be angled toward the sun (south) 10 degrees more than your local latitude. In the central midwest states, in a line through St. Louis, Mo. that is about 40 degrees latitude, this angle would be 50 degrees. Because many roofs are angled closer to 45 degrees, collectors are mounted at this angle, with little loss in efficiency. The basic idea is to have the face of the collectors at right angles to the rays of the sun when it is low on the southern horizon in the winter, which calls for a slightly steeper angle. Collectors for a hot water heater should be closer to 45 degrees to be most efficient all year. In the summer, the sun is almost directly overhead, and the compromise at 45 degrees assures solar heating in the summer as well.

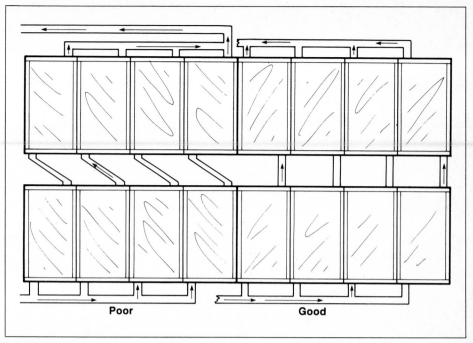

Some hookups of solar collectors take more pipe and fittings than others. The arrangement at right has the shortest length of pipe, but assures pickup of maximum amount of heat.

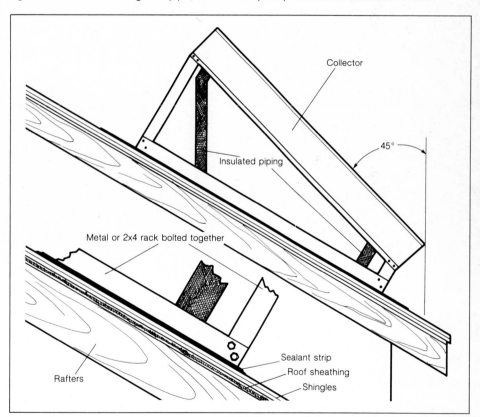

On shallow-pitched roof, mount collector on racks for correct angle. Racks should be of metal or wood, with a full-length base to spread the weight. Single "legs" could damage the roof. Use felt sealer strip between roof surface and rack; run caulking around joints.

Some homeowners have installed their plate collectors on the ground rather than on the roof. There are obvious advantages to roof installation: there are few obstructions, vandalism and damage are less likely, and the "attractive nuisance" problem (children and animals inevitably are attracted to solar plate installations because of the warmth) is avoided.

Flash and seal the edges so that the unit becomes part of the roof. Make sure all connections are watertight. Cover all screws, bolts or fasteners with roofing

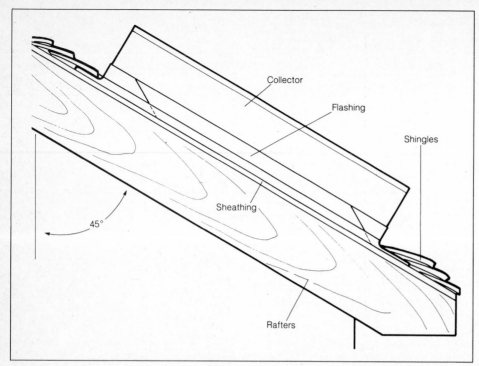

To retrofit, place collectors on roof; use flashing all around. Treat two or more collectors as one, with upper and lower sides as one piece to minimize joints that could leak. This spreads the weight of the collectors more evenly on roof.

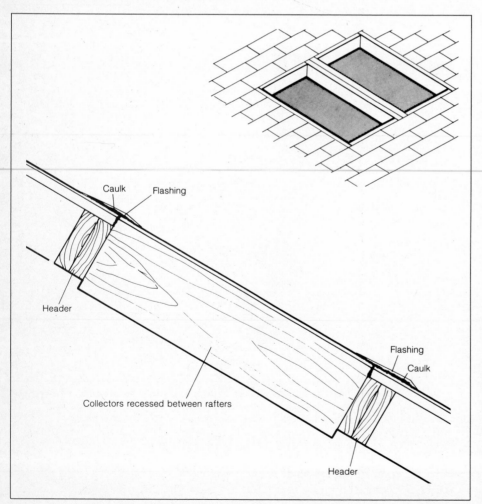

In new construction, or in rebuilt roof, recess collectors between rafters with cover plates flush with roof surface. Fit flashing tightly and caulk all joints. Install doubled headers above and below collector openings. Run rafters full length between collectors.

compound. As an alternative, angle iron brackets could be used in order to raise the collectors above the roof. Seal any screws or bolts driven through the roof covering.

If you build a room addition, the collectors could be built right into the roof and become part of it.

This requires careful planning and insetting of solar collector panels between the rafters, as shown in the illustration (at left, below—note caption details). For sufficient strength, rafters must run full length between collectors.

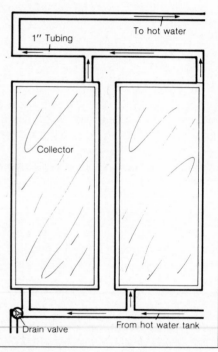

Assemble collectors in "reverse return" layout. This assures equal flow and pressure drop through each collector, so each collector provides maximum heated water. Be sure inlet and outlet of each collector are at opposite corners.

ATTACHING A SOLAR HEATING SYSTEM TO THE WALL OR ROOF

On a south-facing wall or roof, attach this fairly simple structure of wood and corrugated sheet metal.

Building the Frame Using 1x2s, build the frame to the desired size. The installation shown is 4 feet high and 30 feet long. It is positioned under the windows of the second floor of a home. The windows provide passive solar collection; the fan connects to the active part of the system.

Step 1: Adding the Absorber Corrugated steel, which has been painted flat black, is nailed over the 1x2 frame. Do not use flat steel; the corrugations provide additional surface and produce turbulence that assures more efficient heat removal. The corrugated metal can be standard steel roofing, which comes in fairly narrow widths. Lap the edges of the strips, seal the joints with silicone sealer and rivet the edges together. Use blind rivets ("pop" rivets) applied with a rivet gun. The nails should be painted black.

Step 2: Creating the Ducts Cut openings between the wall studs (through the siding) at each end of the collector to create ducts for the inlet of cool air from the house and the exhausting of heated air into the house. The panel will fit over and hide the ducts.

If the house wall is brick you will have to remove bricks, then cut through the sheathing with a keyhole saw. The opening inside is at the bottom of the stud space, the stud space creating a "duct." There is one at each end of the collector.

Step 3: Setting Up Air Flow A small squirrel cage fan that moves 350 cubic feet of air per minute with the aid of a 1/15th horsepower electric motor provides ample circulation of the solar-heated air. A thermostat in the collector should be set to about 75 degrees to turn on the blower. This may take some experimentation; set the thermostat higher and lower to determine the maximum efficiency. In the winter, with the sun low on the horizon, the collector may heat up quickly. In this case, starting the blower at a lower temperature may prevent the collector from overheating. In the summer, the collector should be kept shaded from the overhead sun.

To prevent the circulation of air when heat is not wanted, and to prevent the

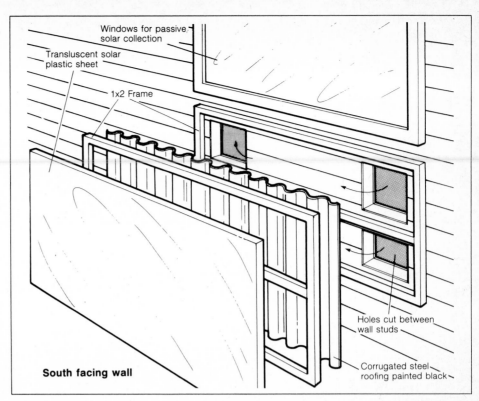

South facing wall

Windows for passive solar collection

Transluscent solar plastic sheet

1x2 Frame

Holes cut between wall studs

Corrugated steel roofing painted black

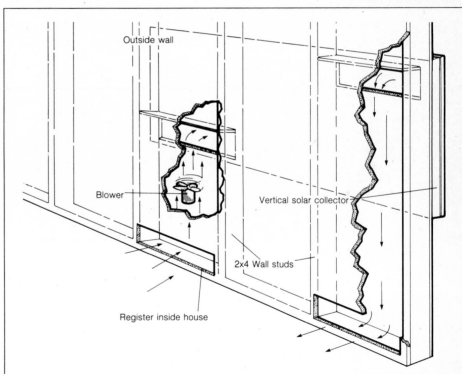

Outside wall

Blower

Register inside house

Vertical solar collector

2x4 Wall studs

To route air for an air-cooled vertical collector, use spaces between 2x4 studs as ducts. Install small blower in the grille of one of the inside registers, or in the duct.

counterflow of warm air from the house when the collector cools, the ducts should have flaps that close by gravity when the fan is not running.

Step 4: Finishing the Unit To finish the collector, build another frame of 1x2s over the corrugated steel panels. Attach fiberglass sheeting, using silicone sealer under the edges. The fiberglass will require a center horizontal strip for support, as will the corrugated steel. Be sure the fiberglass is the type used for solar collection. (Some kinds of fiberglass panels actually have heat screening properties and should not be used.)

ADDING A "MODIFIED TROMBE" WALL

If you are a bit more adventursome and have a south facing exterior brick wall on your home, you can use it as a "Trombe" (pronounced Tromb) wall by building an add-on glass wall. An existing double hung window is used to create upper and lower "vents" that circulate cool air in at the bottom and hot air out at the top. In this case, "out" means into the room you want to heat. On a sunny winter day, the air inside this passive solar collector can reach temperatures higher than 120 degrees F.

Advantages of a "Modified Trombe" Unit

A true Trombe wall consists of a wall of glazing in front of a south-facing wall of masonry, so that the wall absorbs solar heat and later releases it. You can build a lighter, "modified" Trombe wall instead. This creates heat but has no storage capability. Your frame house can be fitted with a modified Trombe wall so that heat is produced in it by the sun during the day, but the wall is "shut off" at night or when the sun is not shining. The "modified Trombe wall" is a lightweight setup that helps capture some of the available solar energy without the need for a heavy mass of material to store the heat. It can be built quickly of 2x4s, at moderate cost and with only rough carpentry skills. Any south-facing wall could be used.

Construction Considerations

The modified Trombe wall can be a "hanging closet" or it can be built on a masonry foundation. A small roof will have to be built at the top of the solar addition and the "ceiling" should have 6 inches of insulation or more, as should the "floor." Side walls also should be heavily insulated.

Glazing can be glass or transluscent plastic. Be sure the plastic is designed for solar transmission, and not for screening out solar heat. If the panels are designed for solar use, there is not much difference between glass and plastic panels as far as solar transmission is concerned.

Closable dampers should be installed in openings near the bottom and top of the solar collector, cut through the wall. Paint the frame fasteners (nails) and the siding of the house a flat black or dark green or some other dark color of non-gloss paint.

The Trombe wall is one method of absorbing and storing heat that is particularly popular when used combined with a greenhouse (see Chapter 6).

A modified Trombe wall is created by adding a hanging closet to a frame home, with lots of glazing. If the addition is over existing double hung sash, sash can be opened at top and bottom to create vents. Cool air enters at bottom, heats and passes back into room at top. Closing window keeps warm air in the room.

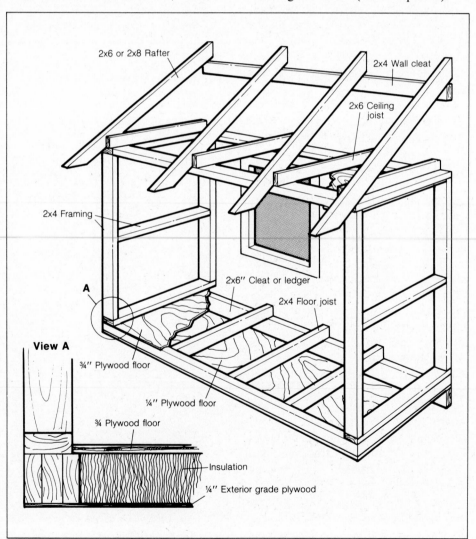

Framing for a "hanging closet" of modified Trombe wall has floor of 2x4s or 2x6s, walls framed with 2x4s, roof framed with 2x4s and 2x6 rafters. Floor is ¾ inch plywood; bottom is ¼ inch plywood. Insulation is placed in the floor. Walls are insulated and covered with ¼ inch plywood, as is ceiling, with more insulation in ceiling than in walls and floor. Bottom rests on a 2x6 ledger. Use bolts or nails, or lag screws and screw anchors.

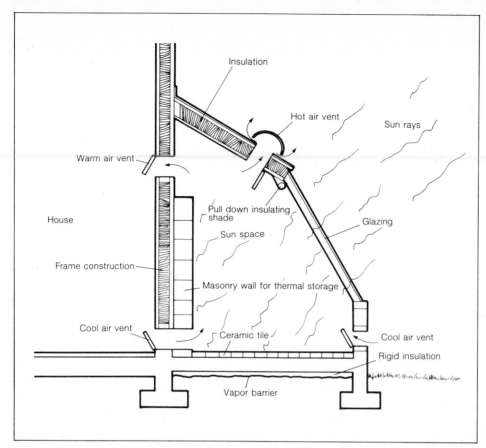

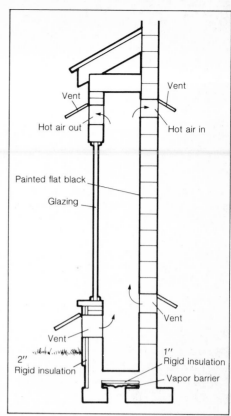

This "sun space" resembles a greenhouse, but is an add-on solar collector. The floor should be masonry; build a masonry wall at the back of the house that has frame construction.

If your home has a south-facing masonry wall, building a glazed structure in front of it results in a Trombe wall that creates and stores solar heat.

A sun room becomes a thermal storage area because of a ceramic tile floor over a concrete base. The floor stores heat during the day, reradiates it into the room at night.

A solar system requires glazing and thermal storage. If you construct a modified Trombe wall, cover the floor below it with tile or other heat-absorbant material.

BUILDING A SOLAR FIREWOOD KILN

The drier your firewood, the less it will weigh and the more heat it will deliver. This means you can buy less wood in order to achieve the same heat value.

Most people try to purchase their firewood supply in the spring, so that it can dry out over the summer and be ready for use by late fall. But this method is not very efficient or dependable. An alternative method is to use a solar kiln to dry out the wood to about 7 percent moisture content. The model for which we give instructions requires about 6 to 8 weeks to dry each load. Since you will be able to buy less wood, the savings will pay back the cost of the kiln in about two years. The only recurring expense will be replacement of the plastic film every one or two years.

The floor of the structure is covered with black plastic film. In front of the wood storage area is placed a screen that has been painted black. The black plastic and screen absorb heat as the sun floods through the front, which is clear plastic. At either side of the roof are open spaces. The air is drawn through the wood, then vented out through the back, pulling moisture out at the same time. Framing members, preferably of long-lasting pressure-treated lumber, are either butted or toenailed with penny spikes for simple construction. The framing and the roof then are painted white. The final step is to attach the clear plastic film to the framing and to the roof.

To prepare the wood for the kiln, split all logs and remove the bark. Check that no termites or insects infest the wood. The size of the load in the kiln will determine the length of time you must keep the wood before using. The kiln will deliver approximately 42,000 BTUs of heat. For quicker drying times, reduce the sizes of the drier loads.

Increasing Efficiency

The dryer illustrated sacrifices some efficiency in its effort to reduce cost. To increase efficiency, install insulation along the back wall. Then substitute double glass for the plastic film. However, addition of glass reduces cost effectiveness and increases the time involved in building the dryer.

To collect more solar energy angle, the 85 sq. ft. roof to equal the latitude; this gives best year-round performance. For better results during winter months, the collector's roof angle should equal the latitude plus ten more degrees. Less efficiency is not really a strong concern; it just means that the firewood will take longer to dry.

Materials List

10 2x4s, 8 ft. long, (pressure-treated) cut to dimensions
3 ⅜ in. plywood sheets (Ext. C-D)
16 penny spikes (1 lb.)
¾ lb. roofing nails
16 ft. black polyethylene sheet
16 ft. clear polyethylene sheet
16 ft. 48-in. wide window screen
1 qt. white latex exterior paint
1 qt. black latex paint

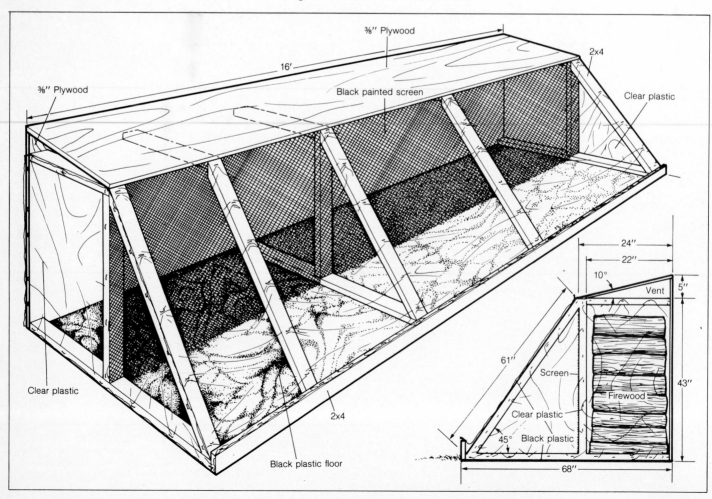

BUILD If you do not feel blower, 1/10 horsepower or smaller, will collector (on a shelf or clamped to a
AIR- up to making a provide the hot air for a workshop room wall), natural thermosiphoning will move
UNIT liquid solar col- or potting shed. the water up into the tank (Thermosi-
, you might want to try a As detailed in the drawing, the back phon systems rely on the "hot water
-cooled unit that can be made is a 4x8-foot sheet of ¼ inch plywood. rises" principle, as does hot air).
t-off beverage cans, a few 1x4 Additionally, as also shown, you can in- For a collector this small, the tank
and some black paint. This basic sert some ⅜ inch copper tubing. This should have a capacity of no more than
is easy to build, is relatively provides hot water. The hot water can 10 or 15 gallons; it should be well in-
and can prove to you beyond be pumped with an equally small motor sulated. The temperature of the water
t the sun can provide hot air or, if a small water tank is positioned probably will not reach a more than 80
water. A small squirrel-cage about 3 feet above the outlet of the solar or 90 degrees.

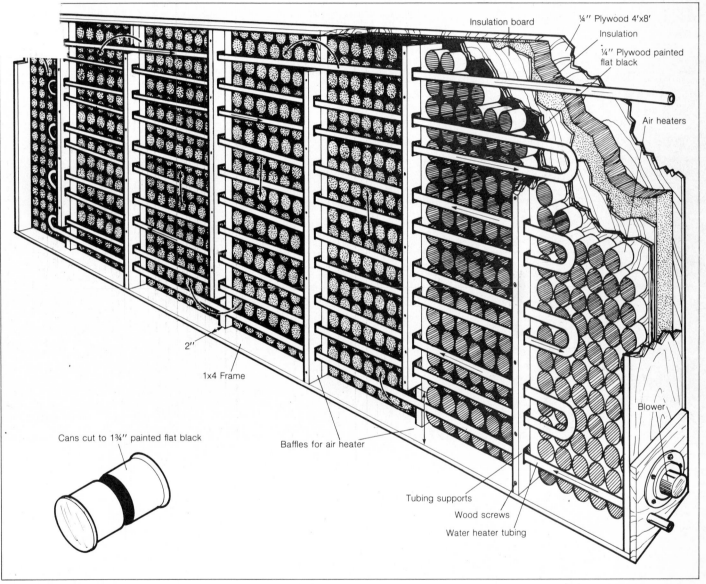

This simple collector, assembled with cut-off food tins and lumber, works without a glazing cover when fins are painted flat black. Plastic sheeting will improve efficiency.

6
Building A Solar Greenhouse

Although a solar greenhouse differs from a conventional greenhouse in its basic function, a casual examination will not reveal much difference. Both types of greenhouse collect solar energy, add living space to a home, can be used for growing vegetables and flowers, and may be attached to the south wall of a building.

The functional difference is that a solar greenhouse will have a means of storing heat and of sending it into the house. A really efficient solar greenhouse will have enough storage capacity so that it will heat itself during nighttime hours and on sunless days. The temperature may drop into the high 30s and lower 40s, but should not approach freezing, which means that cool-weather vegetables can be grown during the winter with no supplementary heat.

During the planning stages of your solar greenhouse, look around your area for homes with greenhouses, especially those that obviously are solar greenhouses. Talk with the owners to learn what they have found to be advantages and disadvantages of certain designs and materials. Solar people usually are delighted to talk about their solar experiments with others who are interested in getting energy from the sun and you can learn much from such people. They very likely will have sources for materials and supplies at the lowest cost, and they might even help you design your solar room.

DOES IT QUALIFY FOR TAX CREDITS?

A solar greenhouse can qualify for a tax credit under some circumstances. However, do not accept the advertising copy that says any greenhouse rates a tax credit. The IRS will give tax credits only to those parts of the solar system that have only one function. If, for example, the masonry wall of your house is being used as a means of storing solar heat collected by a greenhouse built against the house, the wall does not qualify for credit. The reason is that it serves two purposes—it stores heat and functions as a major structural component of the house. The same would be true of heavy drapes you pull across the sliding glass door that closes off the greenhouse at night. The drapes not only insulate the glass doors; they also provide privacy and are part of the decorating scheme of the room.

If the primary purpose of your green-

When adapting a commercially available greenhouse for solar collection, the main addition is a means of storing collected energy.

The brick wall and masonry in this solar greenhouse provide storage for solar heat. A roll-down shade inside provides insulation to keep heat in at night, but out in the summer.

A solar greenhouse need not be small. If large enough, it will provide additional living, dining, or entertainment space.

1 Construction of a large solar greenhouse requires adequate foundation support (for specific instructions, see Chapter 6).
2 The interior woodwork was designed to fulfill structural needs and to tie the room design to that of the rest of the house.
3 Because the solar greenhouse addition was built on a sloping site with a good view, it serves as a leisure and entertaining area.
4 Although this greenhouse also was built large enough to serve as additional living space, the owners chose to restrict its use to plant cultivation.
5 For interior layout, try to keep everything within reach of a central aisle.

3

4

5

1

2

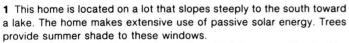

1 This home is located on a lot that slopes steeply to the south toward a lake. The home makes extensive use of passive solar energy. Trees provide summer shade to these windows.

2 These windows give light and warmth to the living room. The double-glazed windows open at the bottom for ventilation. The living room floor is polished wood that absorbs the heat.

3 An indoor swimming pool is on the west of the home. The roof is angled to get maximum benefit from the skylights that provide both light and passive solar heat for the pool water and the enclosing room.

4 Pool water is pumped through this solar collector and heated. The backing of the collector is black as is the rubber tubing that carries the water. The covering is polyethylene sheets.

5 The enclosed swimming pool is heated by solar energy and, in winter, by an auxiliary system that pumps the water through copper pipes embedded in the living room fireplace to give extra heat.

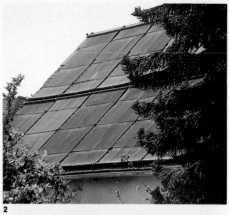

1 Twin solar panels atop this California residence supply all the hot water that the residents require. Supply and return piping leads directly to a hot water storage tank.

2 This residence uses solar energy to heat the water of an outdoor swimming pool. The entire roof is covered with solar collectors. Piping connects the installation and the pool.

3 A passive system, such as this solar pool blanket, requires no extra apparatus to function.

4 In order to achieve efficient heat harvesting, a solar collector should sit at a 45 degree angle, as shown here. Angle iron braces provide firm support.

5 The apparatus shown here has two functions: it pumps hot water to the inground swimming pool and acts as an auxiliary heater on days when the sun does not shine.

1 If possible, choose the size, shape, and materials for the solar flat-plate collectors so they are as unobtrusive as possible.

2 Flat-plate collectors should be mounted on a south-facing roof. For three methods of installation, see Chapter 5.

3 Because the angle of the mounted collector will vary to the slope of the roof, racks are used to hold the unit at a proper angle.

4 If the plate collector is planned and added during construction, rather than later as a retrofit, the roof design can be built to accommodate it.

5 This commercial unit was added without any structural changes to the existing roof.

1, 2 Passive solar heating systems utilize glass walls, preferably south-facing, with exterior walls and interior floors of masonry or other heat-retaining materials.

3 Homeowners who desire to cut energy usage and employ natural means to heat their homes often come up with unusual systems, such as this heat retaining sod-covered roof.

4 Solar greenhouse glazing usually is of glass for looks and permanence, but plastic sheets offer lower cost and are still effective.

5 On the south side of the Sun/Tronic House, flanking the central greenhouse/solarium, are 640 sq. ft. of all-copper collectors.

6 Wood stoves offer backup heat for a solar heating system or for a conventional furnace arrangement. For installation and clearances, see Chapter 10.

house is to grow flowers and/or vegetables rather than to collect solar heat for the house, the greenhouse also is not eligible to tax credit.

Tax credits can be as much as 40% of the cost of eligible solar energy equipment up to a maximum credit of $4,000. The tax credit will expire on December

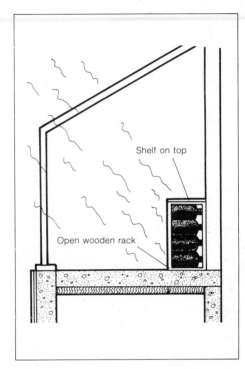

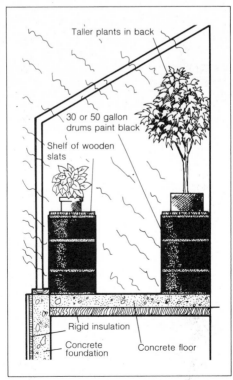

Water storage can be provided by filling 30 or 55 gallon oil drums, plastic gallon jugs or clear glass jugs with water. Paint all containers a dark color and let them receive direct sunlight at least part of the day.

31, 1985, but unused tax credits may be carried over from year to year until December 31, 1987.

Because of the IRS regulations, an add-on solar greenhouse (whether you build it or buy a manufactured unit) has a better chance of qualifying for tax credits than a greenhouse that is built as part of a new house. One heat storage system that will qualify can be of 30- or 55-gallon oil drums painted a dark color, or even a rack filled with empty bleach bottles you fill with water. For most efficient heat collection, the bottles should be painted a dark color. For a more pleasing appearance, you can use clear glass gallon jugs and tint the water a dark color with food dyes.

An underfloor space filled with washed rocks or gravel can be used for solar heat storage. This space, like the bottles or drums, obviously has only one use and definitely qualifies for a tax break.

What Is The IRS Looking For?
The IRS rules specify that a passive solar collector have five elements:

(1) a solar collection area, like south-facing glass or other transluscent material as used in a greenhouse;

(2) an energy absorber such as a mass of masonry or water that is heated by solar energy;

(3) a mass of material that is large enough to store and deliver adequate amount of solar heat for the relative size of the house. The preceding requirement would seem to be the same, but the IRS might think differently.

(4) a means of releasing the heat from the storage mass by radiant or convective means;

(5) heat regulating devices (awnings, insulated drapes or automatic vents) that control the amount of heat admitted to the house during the day and reduce heat loss at night.

With these kinds of regulations, homeowners can become confused and make claims that are not allowable according to the law.

COMMERCIAL UNITS
Commercial greenhouses come in several configurations. There is one type with a slanting south-facing wall that joins the angled roof by means of a curved section of glass. Another type has the angled wall joining the roof at a sharp angle. Still another has a vertical south-facing wall and an angled roof. All of these types (except for the greenhouses designed specifically as solar collectors) have glazed east and west end walls. The roofs are also glazed.

The instructions that come with a commercial greenhouse should tell you how to fasten the unit to the foundation and to the house.

Ventilation Needs
The unit should come with vents in the roof and near the bottom of the greenhouse to create circulation during the summer, when a greenhouse tends to overheat. Cool air, entering the lower vents is heated and moves up out the roof vents.

Solar greenhouse does not need full glazing in end walls, but insulated frame walls instead. This solar greenhouse has transluscent roof glazing to cut down summertime heating.

CONVERTING A COMMERCIAL GREENHOUSE TO SOLAR USE

If you purchase a commercial unit and want to use it as a solar collector, we recommend that you frame in and insulate part of the roof and at least part of the end walls. Maximum heat harvesting is done from about 10 a.m. until about 3 p.m., when the sun will send its rays almost directly into a south-facing greenhouse. Insulated end walls help retain the heat provided by the sun during the harvesting hours.

Depth of Wall Enclosures To determine how much of the end walls to close in, check in the morning to see where the rays of the sun start to contact the back wall at about 10 a.m. At this time, hold a solid sheet of cardboard flat at the outer corner against the east wall

You can buy a "kit" for a solar greenhouse. This one has an outside entrance and wooden siding to provide thermal storage.

This solar greenhouse sets on a foundation that is almost flush with ground. Exterior shades protect the inside from the summer sun.

framing. Slide the sheet back towards the house until the sun's rays are no longer shaded by the outermost edge of the cardboard and instead hit the house wall. Note the location of the outermost side of the cardboard. Build the east wall out this far. Repeat the procedure on the west wall framing at about 3 p.m.

We'll have to admit that this is not very scientific, and you probably can build the wall out farther, or even make it shorter, and still have a successful heat-producing solar greenhouse.

Depth of Roof It is not uncommon for a greenhouse to become too warm during the summer. In order to shade out direct overhead sunlight during the hot summer months, plan for the maximum shading as the sun is directly overhead on June 21, when the sun is directly overhead. Build the roof out far enough to achieve this shading. In addition, you can install split bamboo or other types of shades that will extend the shadow line in the greenhouse during July and August when it is very hot and you want the minimum of sun in the greenhouse.

Step 1: Enclosing the Walls First, position the large sheets of rigid insulation board over the wall areas that you intend to enclose. Use nails to tack the sheets to the greenhouse framing. Then install siding or other outer sheathing to match the house. Fasten the siding to the metal or wood frame with 4 inch screws to hold both the sheathing and the insulation in position.

Step 2: Closing the Roof Again, sandwich the insulation board between the greenhouse framing and ¾ inch plywood sheets that function as the roof decking. Fasten the two in place with 3½ inch screws. Tack a sheet of felt underlayment over the plywood deck. Install a strip of metal flashing to cover the joint between the greenhouse and the existing roof. Then shingle the roof to match the existing roof.

CONSTRUCTION GUIDELINES— HOMEMADE UNITS

A homemade unit must have all the features of a commercial unit.

Orientation

A solar greenhouse must meet two conditions: the house must have a wall facing within 30 degrees of true south; and nothing must be allowed to shade the greenhouse when the sun moves to its lowest point on the horizon in the winter. Leaf-bearing (deciduous) trees are all right, since they lose leaves in the winter and their branches create only minor shading.

As a rule of thumb, the farther north you are located, the closer to vertical the south facing wall of the greenhouse should be in order to catch maximum sunlight. In any location the closer to vertical the wall is, the more usable interior space you will have.

Storage Mass

To figure how much storage mass your greenhouse requires, use this rule of thumb: for every square foot of glazing, include either 75 pounds of rock or masonry, or two cubic feet of water. In other words, for masonry you require three times as much in volume, and five times as much in weight, compared to water storage.

Quite obviously, water will take up much less space and require much less structural reinforcement than rocks or masonry. However, some people prefer masonry or rocks because water containers can leak, or water can "turn bad" because of contaminants in it. Water used for heat storage should be treated with an algecide or a combination algecide/purifier. These materials can be purchased from suppliers of solar devices. If there are none locally, you can purchase the chemicals by mail order.

Doorway Locations

Ask yourself: will you want an entrance to the greenhouse from the house or will you be satisfied with an outside entrance. It is much more difficult to move

This installation sits in the L-shape of a house. Open windows give ventilation and circulation.

the heat into the house when there is no entrance between the two structures. Any outside door in the greenhouse immediately creates problems with air infiltration.

If the back wall of the greenhouse includes a house window, open it up to make a door. Unless you have had experience with wall framing, it would be best to have a professional carpenter or contractor do this job. You will be working with an outside, load-bearing wall and will have to support the ceiling while the opening is made, a header installed and the opening properly framed. Be sure to obtain the doors before you make

Ceramic tile-covered concrete slab provides heat storage. When doors are open, light, heat and humidity flood into the house.

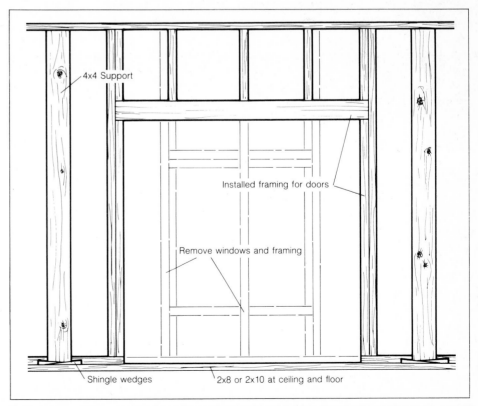

4x4 Support

Installed framing for doors

Remove windows and framing

Shingle wedges

2x8 or 2x10 at ceiling and floor

Before you cut through a load-bearing wall, support the ceiling. Frame the opening with double headers. The size of the opening depends on the size of the doors.

the opening, as doors vary in size as does the required rough opening.

Door Styles Sliding glass doors are a popular installation because they enable people to see into the greenhouse even when the doors are closed. To minimize the loss of heat by conduction and infiltration, be sure the doors have insulating glass and weatherstripping.

French doors offer much the same advantages as sliding doors. In addition, they are more easily weatherstripped, so there is even less air infiltration than with sliding doors. French doors do have a disadvantage. They require floor space for the doors to swing open. Because the greenhouse generally is not large, the doors open into the house, which can cause problems with furniture placement. Where space is restricted, sliding glass doors will be more desirable.

Work Schedule Make the opening for the entrance into the house after the greenhouse is partly closed in so weather is not a problem. There must also be an opening in the greenhouse so you can work from the outside as well as the inside. Lumber, tools and the like can be passed through an opening in the greenhouse so the mess is not tracked through the house.

Doors should have insulated double glazing, as there will be times when you want to seal off the greenhouse from the main house. This is especially true in the summer when there will be times when the greenhouse overheats and even venting cannot keep down the temperature.

Foundation Walls

The glass panes in a greenhouse either come right down to the ground or end at the top of a wall that projects 3 or 4 feet above ground. For a solar greenhouse, the projecting wall makes sense for a couple of reasons. First, the wall becomes a mass for the storage of heat. Second, the glass or other glazing is above the ground and less subject to damage from lawnmowers and other lawn and garden equipment. If you build a commercial greenhouse with a slanting wall, the short masonry wall also assures more usable floor space.

Ventilation

If you build your own greenhouse from scratch, you must provide for ventilation vents. The vents in a solid roof are made

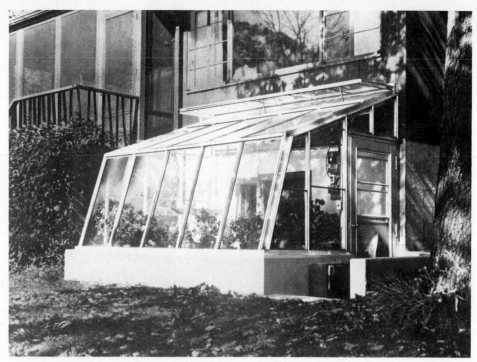

This installation takes advantage of a heavy concrete block foundation as part of the storage facilities.

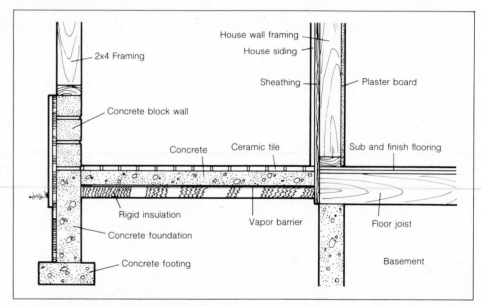

The best arrangement has the walking floor of solar greenhouse flush with house floor, whether the house has a basement, or is on slab or over crawl space.

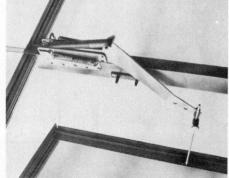

Ventilation system automatically opens or closes when the temperature inside a greenhouse reaches a certain level. These rely on chemical reaction, not electricity, to operate.

like trap doors. Gaskets around the edges seal the doors so they are waterproof. You can open the vents by hand or use heat actuated devices. These temperature-sensing devices will lift 40 or 50 pounds when the heat inside reaches a pre-set level.

If you build your own greenhouse you may want to cast openings in a wall that projects above ground, or leave openings if you lay a concrete block wall. The doors should be hinged over the openings, and faced with rigid insulation. These vents in the wall generally are manually operated, since they are easier to reach than those in the ceiling, but they can also be actuated by a temperature sensing device like those for the ceiling vents.

One problem that might occur with heat activated devices is that the vents in the ceiling (roof) might open before the ones near floor level because of the higher temperatures near the ceiling. To solve this problem, set the floor level vents to open at a lower temperature than those in the ceiling.

Various kinds of vents can be purchased from makers of greenhouses, and you should obtain catalogs from several of the manufacturers to compare.

MATERIALS NEEDED
Framing
Because of the high humidity in a greenhouse, and because the wall framing is near the ground, pressure-treated lumber or redwood is recommended for the wall framing. The cost of the treated lumber is generally only a bit higher than untreated lumber. Roof framing may be of standard lumber, as the roof structure drains rapidly and the heat of the greenhouse also keeps the wood dry. However, in humid regions we recommend use of treated lumber for the roof framing also.

Keep the framing for the greenhouse as small as possible to maximize the area of glazing, but remember that structural strength also is important. While 2x4 framing is recommended for the average size greenhouse, a structure that is fairly shallow (short from the house wall to the greenhouse wall) can be built with 1x2 lumber. About every three or four frames use a 2x2 to assure adequate strength.

Glazing
A solar greenhouse should have double glazing. This will reduce the transmission of energy through the glazing somewhat, but it also prevents the outflow of energy found with single glazing. Glass is the most efficient material for glazing, although some plastics do almost as well. If vandalism might be a problem in your area, use plastic on the outside of the framing and glass on the inside. Otherwise reverse the positions of glass and plastic. Use plastic panels that are formulated to allow solar energy to pass. These cost somewhat more than the sheets of plastic used for patio covers, but the patio material may screen out 60 percent or more of the heat energy and will completely defeat the purpose of a solar greenhouse.

Size The size of the panels should not exceed 24 or 30 inches, since beyond that you will have problems even with double strength glass. Double glazed plastic panels are stronger than glass and will span a wider framing, but these are somewhat more expensive than glass. Choose a design that will deliver the most strength and glazing area with a minimum of framing and the least possible cost.

Convenience Be sure to make the inner pieces of glazing removable so you can clean the inside surface of the outer glazing. At the same time, seal the inner panels of glazing so that there is no air movement in and out of the space between the inner and outer glazing.

Select a glass or plastic size that is standard so that you don't have to pay extra for custom made material. In the event that a panel of glazing is broken, you want material that is readily available to permit quickly replacing the broken pane. It's a smart idea to have two or three spare panels stored near the greenhouse so you can quickly reach the panels to repair the greenhouse before too much damage occurs.

Insulation
The end walls of a solar greenhouse do

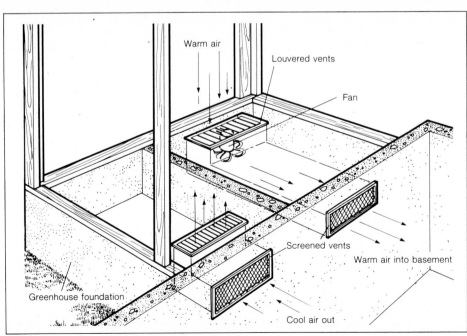

Heat can be ducted down under a greenhouse and into the house basement. The foundation under greenhouse can be divided so there is supply and return of warmed and cooled air.

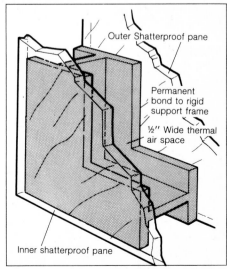

A solar greenhouse should have double glazing to prevent heat loss through the glass. Some manufactured units provide this.

not necessarily have to be glazed. Instead, they can be solid masonry with insulation on the outside, or stud wall construction. The masonry wall will require a covering of some kind over the insulation to protect it.

Insulation goes between the studs in a frame wall. To increase the insulation, assemble the stud walls from 2x6s rather than 2x4s. This permits installing insulation 5½ inches thick, which is two inches thicker than the insulation that can be installed in a 2x4 wall.

Fasteners

Because moisture is a problem both inside and outside the greenhouse, use either galvanized fasteners or aluminum ones. Nails, more practical than screws for most of the assembly, make the job go faster. Use screws wherever you want to create a stronger joint than you can get with nails. The joint between the roof and wall of a greenhouse might be a place where additional strength is required. The angle there produces greater stress than a complete right angle.

PLANNING AND CREATING THE FOUNDATION

Whether you build your own solar greenhouse or buy one from a manufacturer and then assemble it, you must provide a solid foundation. Not only is this good construction practice, but almost every building code requires it. Either pour a concrete footing below the frost line and then build a concrete block wall on top of it, or pour the footing and wall as one solid mass of concrete. Check local building codes for instructions on how the greenhouse foundation should meet the house foundation. Generally, there should be no positive connection between the two foundations, so that the settling or movement of one will not affect the other. If possible, plan the position of the greenhouse so that the framing aligns with the house framing.

This will greatly simplify the task of attaching greenhouse to the wall.

CONCRETE FOOTING WITH BLOCK FOUNDATION

The bottom of a footing should be below the frost line, and your local building code will specify the dimensions. At the least, the depth will have to be equal to the width of the foundation wall. For 8-inch concrete block, the footing will have to be from 12 to 16 inches wide and at least as deep as the block, or 8 inches.

Step 1: Pouring the Footing First, dig a trench that is wider than the footing will be. Set up forms into which the concrete will be poured. These forms should be constructed of 2x4s (or wider) and well braced, as shown. Oil the forms with motor oil for easy removal later.

Be sure that the forms are level and plumb before you place the concrete. Pour the concrete into the footing forms, screed level. After half an hour or so, when the concrete holds a thumbprint, smooth with a wooden float.

Step 2: Laying the Courses When the concrete has set for 24 hours, you can remove the forms and lay the concrete block. The first course is set in a thick layer of mortar to allow for any necessary leveling of the course. Mortar lines between courses and between blocks are made about ⅜ inches thick.

Bring this wall up to whatever level you desire. Usually the courses are built up to the floor level of the house, although there can be a step up or down into the greenhouse.

Laying a plumb and level wall of block takes care and some skill. Every course must be checked for level. Build corner leads first; then fill in, working from the corners toward the middle. Set anchor bolts (see below) in the top course of block, after filling the block voids with mortar. The anchor bolts hold the sill plate at the base of the greenhouse framing.

Step 3: Insulating the Foundation. Once the mortar has set, adhere rigid insulation to the block wall with mastic. Then backfill the soil against the wall. This soil will hold the insulation against the wall, even if the mastic later degrades and loses its adhesive properties. Later, the insulation above ground will have to be covered with fiberbond, siding or other protection. Otherwise the board will degrade due to weathering and sunlight.

POURING A CONCRETE FOOTING AND FOUNDATION

A second method of making the foundation is to pour the footing so that it has a groove, or "key," in it. Later, be-

The key in a footing is created by pushing 2x4 into fresh concrete. When foundation wall is poured, it fills the key and is "locked" so it will not move sideways under pressure. In clay or other heavy soil, foundation can be dug in earth with soil creating the form. Any portion of wall to be above ground must be formed. Use tie wires between the plywood forms to keep them from spreading.

Labels on illustration: 2x4 Brace, 2x4 Frame, Wire ties, Foundation wall, Diagonal brace, Exterior grade plywood, Trench footing, Keyed footing

fore the pour that creates the foundation wall, the key is removed. Forms are then set for the foundation walls and the concrete set inside. The concrete of the wall fills the key groove to lock the wall and the footing together.

In some areas, where the soil is mostly clay, you can dig the form for the footing and wall right in the soil. The footing is formed by digging out the bottom of the trench in a sort of "bell shape" so that the part of the trench for the footing is wider than the part for the wall. However, if any of the wall is to project above the ground, forms must be set up to contain the concrete.

Step 1: Creating a Keyed Footing Set the forms for the footing and place the concrete. As the concrete sets up, prepare a 2x4 that is as long as the footing. Bevel one narrow face into a squared-off V-shape. Cover this beveled edge with a generous coating of motor oil. After the concrete has been screeded off, press this key into the plastic concrete. Let the concrete set up about 24 hours or more. Remove the beveled board.

Step 2: Preparing the Foundation Wall Place the forms for the foundation walls. These too must be oiled for easy removal and securely braced. Position 1 or 2 inches of rigid insulation so that it will be on the outside of the wall when the concrete hardens. The insulation should reach down to the top of the footing and extend up the full height of the wall.

Step 3: Pouring the Concrete Place the concrete and screed it level. Position the anchor bolts (see below) while the concrete is still plastic. Remove the forms after 24 hours. Let the concrete cure at least seven days.

Curing Concrete Usually the surface of the concrete must be kept moist during this period. Spray the wall periodically with a fine mist from a garden hose. A burlap cover will help hold in the moisture. A damp cure is necessary for concrete to develop its full strength.

INSTALLING THE ANCHOR BOLTS

The anchor bolts in the foundation are positioned between studs or other vertical wall framing, and not under them. Mark off a 2x4 or 2x6 sill plate to show the locations of the vertical framing members. These can be spaced 16 or 24 inches on center as in standard wall framing, or even wider, depending on the design of your greenhouse. Commercial greenhouses may have the framing spaced 30 or more inches apart to accept panes of glazing that are that wide.

Bore the holes for the anchor bolts between the positions of the members. Then cover one side of the plate with motor oil. Place this face on the foundation. Push the anchor bolts down through the holes. Turn the nuts on beforehand so all the anchor bolts are recessed the same depth and project a thread or two above the nuts.

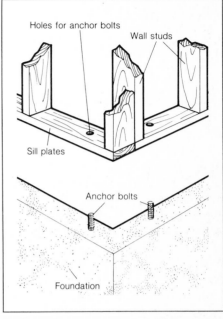

Anchor bolts are set in the last (top) course of foundation wall. Mark locations of 2x4 studs onto sill plate. Use plate to locate anchor bolts which should not fall under a stud.

LAYING A GROUND-LEVEL FLOOR

This floor will be either the walking surface or a subfloor of the greenhouse, depending upon your design. Because a greenhouse is a relatively lightweight structure, no reinforcing is required in the footing and foundation walls, unless local code requires it. The floor can be reinforced if you wish, to minimize any cracking in later years.

Step 1: Preparing the Base Solidly tamp the ground inside the walls of the foundation. Then spread an inch or two of sand on the ground; level this with a dragboard. Cover the sand with a layer of 4 to 6 mil polyethylene for a vapor barrier. Over this, place a 1- or 2-inch layer of rigid insulation. Do not puncture the vapor barrier or the insulation as you work.

Step 2: Adding Reinforcement The reinforcement used is of welded mesh. This must fall parallel to the ground, about half-way through the slab. Before you place any of the concrete, set small rocks on top of the insulation. This will keep the reinforcement mesh from sinking down when the concrete is placed over it. Complete about half of the slab. Then position the mesh and finish the rest of the pour.

Step 3: Placing the Concrete The concrete floor will be contained inside the foundation walls, so you require no side forms. Place an asphalt-impregnated isolation strip across the surface of the house foundation.

Because of the width of the greenhouse, you may need to make the pour in two steps. One pour would be too difficult to screed level. A two-part pour requires that you separate the pours with a temporary stopboard down the middle of the floor, running from the house foundation to the greenhouse outer wall. Use a 4x4 for this stop board. Its weight is great enough to hold it in

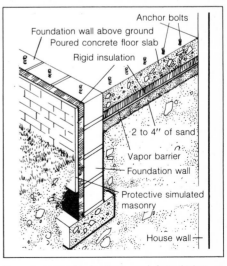

Pour a concrete footing, and foundation, or build foundation of concrete block. Insulate under floor and outside footing, as well as foundation and outside wall if above ground.

place as you pour the concrete without installing side braces that would puncture the vapor barrier. With a helper, screed the first section of the floor. After about half an hour, remove the 4x4 and place the rest of the concrete. Screed this section. Let the concrete cure for at least seven days.

ALTERNATIVE METHOD: FLOOR JOIST SHELF

One way to provide built-in support for floor joists is to form up a poured foundation so a "shelf" is created on the inner top edges. For a concrete block foundation, the "shelf" is created by using blocks 4 inches thick on top of the ordinary blocks that are 8 inches thick, so a width of 4 inches is left on the inner

top edge of the wall. Anchor bolts are set in either the block or concrete wall, and are positioned in the center of the "shelf" to permit bolting a 2 x 4 sill

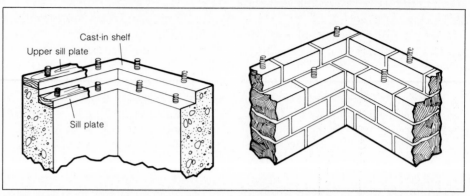

Another way to position floor joists is to form or make a shelf half the width of the foundation wall. Sill plate bolts to the shelf to which joists are spiked; anchor bolts in top of wall hold wall framing.

plate (a 2 x 4 is 3½ inches wide) to the shelf. The floor joists then are toe nailed to the sill plate. A plywood or strip floor can then be nailed to the joists.

INSTALLING A WALKING FLOOR

If the concrete floor is not to be the walking floor of your greenhouse, there are three optional floor designs that you can choose from. All will provide a floor that is level with the existing floor of your house. The floor of the greenhouse can be a reinforced concrete slab, a grid of closely spaced redwood or pressure-treated strips, or sheets of exterior grade plywood.

A Poured Concrete Floor

The poured concrete floor is a job for a professional, since the floor is poured over metal forms that must be adequately supported until the concrete

with its reinforcing bars, sets solidly. The metal forms then become the "ceiling" of the storage space. The slab is poured so that its edges rest on the foundation walls and are flush with the outsides of the foundation walls. It is necessary to have a concrete block or poured concrete wall against the house foundation to support the inner edge of the floor slab. Openings are cast in the floor to provide an air inlet and an outlet. A blower forces heated air from the greenhouse down into the space under the floor where the heat is absorbed by

containers of water or by rocks. The storage material must be installed before the floor is poured.

A WOODEN SLAT FLOOR

A slat floor allows a natural exchange of air from the greenhouse into the storage material under the floor. The joists are nested inside the foundation walls.

Step 1: Constructing the Framing
First, remove the sill plate and construct the wall framing as discussed below. Fasten the framing to the foundation wall.

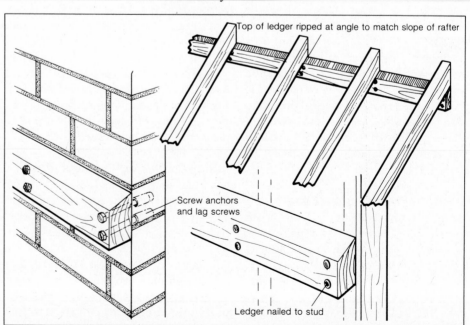

Floor of reinforced concrete requires professional work. Corrugated metal form remains after concrete has set. Reinforcing bars run length and width of floor.

Use masonry screw anchors and lag screws to fasten a ledger to a masonry wall. To fasten to a frame wall, use lag screws. If the ledger meets the house roof line, rip the ledger to match the angle of the greenhouse rafters.

Step 2: Installing the Ledger Install a 2x4 or a 2x6 ledger board to the house. Spike the ledger to a frame house; for a brick or masonry house, use lead or fiber screw anchors and lag screws.

Step 3: Adding the Joists Then, at 16 or 24 inches on center, set joists with joist hangers to the ledger board and the sill plate. The joists also can be toe nailed to the sill plate.

Step 4: Spacing the Wood Slats Lay 2x4s flat across the joists. There should be a space of no more than ⅛ inch between each of the boards. This spacing allows circulation, but prevents heels from getting caught in the openings. In addition, only the smallest of items can fall through to the storage area.

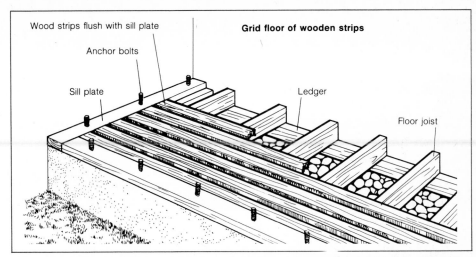

Floor joists can be supported on ledger strips fastened to the insides of the walls. Use either fiber or lead screw anchors to accept heavy lag screws. Since foundation wall is 6 to 8 inches wide, a 2x4 sill plate leaves a small space between sill plate and joists. A wider sill plate, to completely cover the top of the foundation wall, is even better.

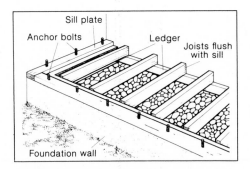

SHEET PLYWOOD FLOOR

Step 1: Installing the Floor Support This floor framing sits on top of the foundation. Leave the sill plate in position on the foundation. Toenail joists to the sill plate and the ledgerboard; cut the joists to allow for a header joist that encloses the outer face of the greenhouse. End joists fall over the sill plates along the two sides. Install 6 to 12 inches of insulation in the spaces along the header joists and along the inside of the end joists.

Step 2: Laying the Sheets The standard thickness for a plywood flooring is ⅝ or ¾ inch Exterior grade. In diagonal corners of the floor, cut openings for an inlet and an outlet for heated air from the greenhouse. A small blower can be attached to a plywood frame that lifts in and out of either the outlet or inlet in the floor. This makes the blower easy to remove for maintenance or replacement.

Step 3: Constructing the Framing Build the framing as discussed below. In this method a sole plate takes the place of the sill plate at the base of the wall frame.

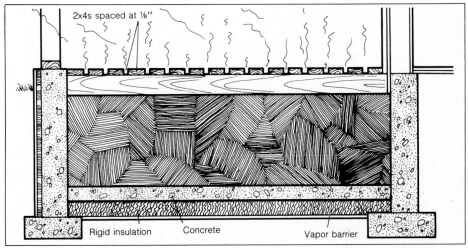

For greenhouse floor of spaced 2x4s, place them flat. Air circulates down to heat storage medium in "basement" under greenhouse. Insulate underneath outside the foundation walls.

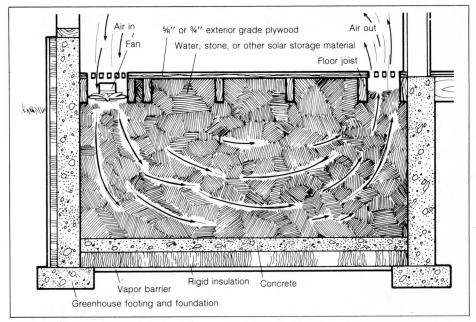

Floor of greenhouse can be ⅝ or ¾ inch plywood (exterior grade) nailed to 2 inch floor joists. Fan blows heat down into space under floor to heat up thermal storage medium.

ADDING THE GREENHOUSE FRAMING AND GLAZING

Before you begin, remove the sill plate from the foundation. Fit a sealing strip between the sill and the foundation. The strip assures an airtight and watertight joint.

Step 1: Cutting the Framing To create grooves in the framing to hold the glazing, cut rabbets that are ¼ inch wide and ⅜ inch deep in the framing members used for the walls and the roof. These will be either 1x2s or 2x4s. An alternative method is to use 4d nails to nail on strips of ¼ or ⁵⁄₁₆ inch lattice stock. Both methods are shown here. The rabbets will create open spaces where the cross pieces join the longer longitudinals. Fill these spaces with short blocks, or cut the crosspieces with a "lip" on each end to fit in the rabbets of the longitudinals.

Step 2: Painting Cut and fit all pieces first; then paint all the pieces before doing the actual assembling. This assures that even surfaces hidden in joints would be protected against the high humidity in a greenhouse. Paint the rabbets where the glazing fits before you install the glazing. This seals the wood; otherwise it would soak up the oil in the glazing compound, which would then fail. Also paint the sill plate, including the underside that bolts to the foundation. Then nail the construction together and raise it into position.

Step 3: Installing the Roof At the position where the roof of the greenhouse meets the siding, the roof of the greenhouse is supported by a ledger board, just as is the floor of the greenhouse. Attach the ledger with lag screws or 3½ inch masonry nails, as shown. If you are attaching the greenhouse roof to the house roof, the ledger will have to be ripped to match the angle of the

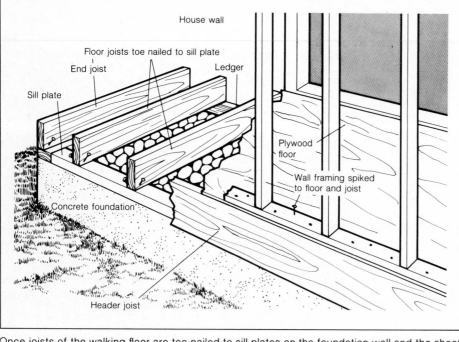

Once joists of the walking floor are toe-nailed to sill plates on the foundation wall and the sheet floor is laid, spike the framing to the floor and header joist.

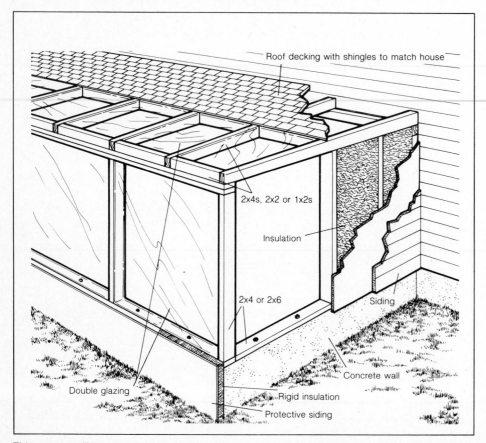

This cutaway illustrates the layers of construction of a solar greenhouse. Glazing covers only part of the roof to prevent overheating in the summer. This side wall must have both insulation and framing panels. Cover the walls with siding to match the house.

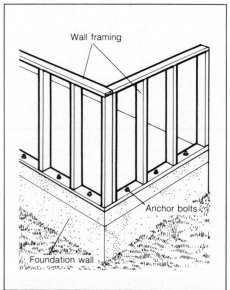

When concrete in foundation wall has set two or three days, remove sill plate. Assemble wall framing on flat surface to permit nailing through top and bottom plates into 2x4 studs for quick and easy assembly.

greenhouse rafters. Position the roof framing and flashing as shown. Then install roof decking as desired. Shingle the decking to match that of the existing roof.

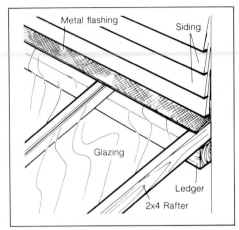

Attach the framing to the side of the house by means of a ledger board. Lay in a strip of metal flashing to protect the seam.

Step 4: Installing the Glazing

The outer greenhouse glazing is held in place with ordinary glazing points and caulking, as is any window installation. To ease the installation of the inner panes so that they will be easy to remove as needed for cleaning, install strips of ¼-inch wood cleats along the rabbeting. Hold the cleats in position with 4d nails every 8 inches or so. This will create an

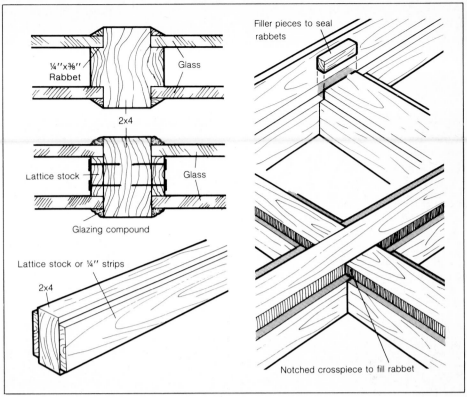

Framing for owner-built greenhouse is straight-grained redwood. Framing can be rabbeted for glass or rabbets can be created by nailing on strips of lattice stock. Install outer glazing with glazing points and caulk.

attractive, flexible installation, which is held securely in place. The caulking holding the outer glazing should be gradually beveled to prevent snow and ice buildup. For installation specifica-

tions, refer to the illustrations shown above at right and center.

COPING WITH PROBLEMS IN THE SYSTEM

No matter how much care and planning you exercise, you may still have problems with the greenhouse. The following suggestions will help solve some common difficulties.

Overheating

If you find the greenhouse overheating, increase the size of the vents in the roof and the wall. If the vents are large enough, the flow of air in at the bottom and out the roof vents will create a "chimney" effect to reduce the overheating.

Providing Supplemental Heat

At night and during days when there is little or no sun, the solar greenhouse may need supplemental heat. This can be provided by simply opening the door between the house and greenhouse so

there is a flow of warm air into the greenhouse to keep your plants from freezing. Many cool weather plants will tolerate temperatures as low as 40° F.

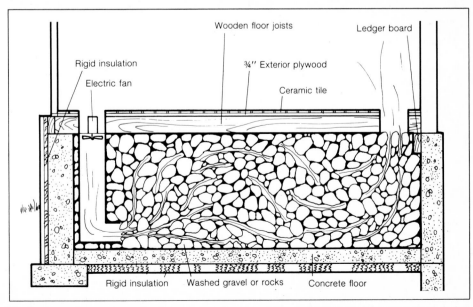

Washed gravel or rocks under floor are used to store heat. A fan that blows heated air in during sunlight hours reverses to blow heated air back into greenhouse when sun is hidden.

Insufficient Heat Storage

The big difference between a regular greenhouse and a solar greenhouse, as previously described, is that a solar greenhouse has storage for the solar heat. If your solar greenhouse cools down quickly, it indicates you need more storage mass. Or, it might mean that the stored heat is not being transferred into the greenhouse, or at least not transferred rapidly enough.

Blower Requirements If the storage is under the floor and a blower is used to push the heated air down into rocks, water or other storage material, the blower must be reversed when the heat is to be supplied to the greenhouse. Natural thermosiphoning will not move the heat into the greenhouse rapidly enough to offset the loss of heat through the glazing.

Some homeowners have reversible blowers in the wall between the house and solar greenhouse in addition to doors that open to allow natural movement of the warmed air from the greenhouse into the house. The fan must be reversible so that when the weather is extremely cold the fan will pass warm air out to the greenhouse, if necessary.

Thermal Insulation To prevent heat loss, install some sort of thermal insulation—insulating blankets or oversized shades—inside the glass or other glazing. Some homeowners fit blocks of rigid insulation into each pane of glazing, although this takes a lot of time and is not too efficient.

One material that should be investigated is a product that looks much like clear plastic and is fitted between the panes of glazing. This thermal material is transparent to solar rays, but reflects heat back inside the greenhouse. Claims of a 30% increase in efficiency are made.

THE KUCKO PROJECT: A "LIVE-IN SOLAR COLLECTOR"

It is obviously more difficult to add a solar greenhouse to an existing home than to include one in the plans for a new one. However, the project is not an impossible task. Since 1978, Gary Kucko, a Wisconsin architect, has been involved in renovating an 80-year-old home for solar capability. As he did so, he strived to maintain the unique character of the old home. Some changes did not affect the home's appearance; others did. The goal was to make the new structure fit in with the old.

The beginning changes were not radical ones. He replaced all the windows with triple glazed units. He then had cellulose insulation blown in to fill the cavity of the exterior walls. The old foundation was made of rock rubble, so Mr. Kucko replaced it with insulated concrete block. He installed an air-lock rear entry and a supplemental wood

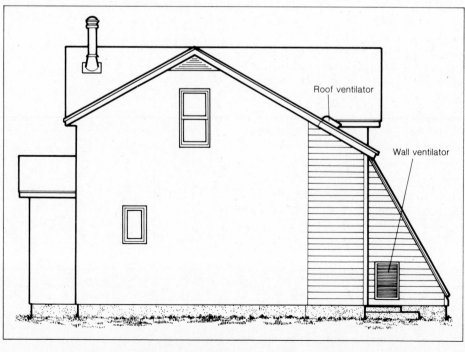

The return-air underground PBC duct is placed before backfilling. Fiberbond coats the 2-inch styrofoam in areas exposed to sunlight.

An 8-inch concrete block foundation will have the 2-inch of styrofoam applied with an adhesive to the exterior of the block wall.

Roof ventilator

Wall ventilator

The floor area is backfilled; 6 inches of sand placed on top of the backfill, and redwood sill plates installed. Siding covers upper layer of styrofoam.

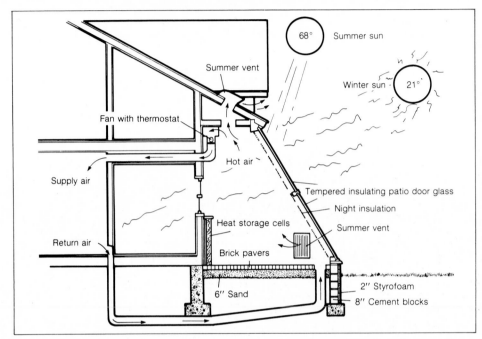

This cutaway shows the air flow and heat transmission patterns designed into the greenhouse. The summer vents are important features that prevent overheating.

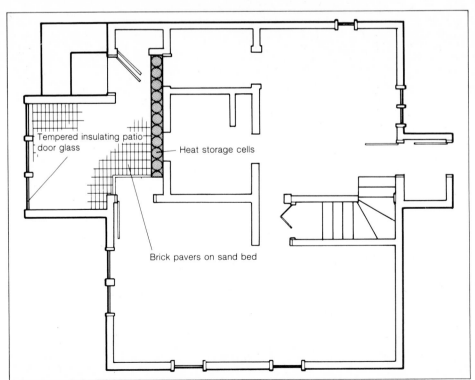

The greenhouse is to be an integral part of the structure of the house itself, so the entrance to the house proper is through the lower part of the greenhouse.

Shown in place are redwood framing for the sloped glazing, sidewall framing for the collector area, and roof covering and sub fascia.

Shown is lower wall vent with insulated shutter in open position. Shutter has 1½ inch Thermax core sandwiched between plywood. Brick floor is of 4x4x12 bricks, not yet grouted.

Views from the southwest and southeast prior to applying the siding. Lower wall vents are for summer ventilation.

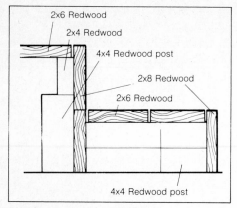

The steps to the front door are a simple construction of 2x4 and 4x4 blocking over which is attached 2x6 redwood lumber.

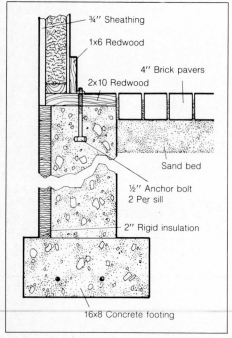

The footing reaches below the frost line. Sand bed and pavers provide thermal storage.

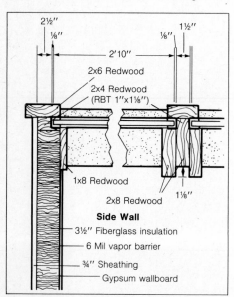

An overhead cutaway shows the glazing and the framing of the side and front walls.

stove. Skylights were included for natural light and solar gain.

He then added a solar greenhouse, which he describes as "A Live-in Solar Collector," for it combines the relocation of the front door, an airlock entry,

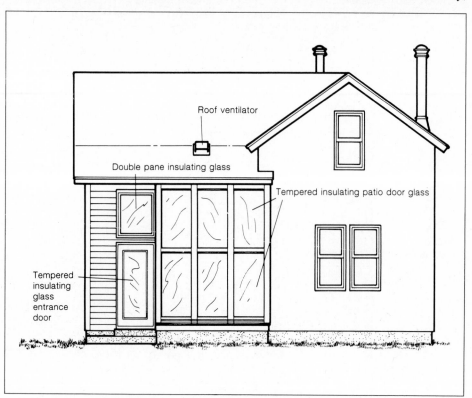

The front facade of the house must be planned carefully so that the greenhouse does not intrude upon the design of the house but appears to be a natural extension of it.

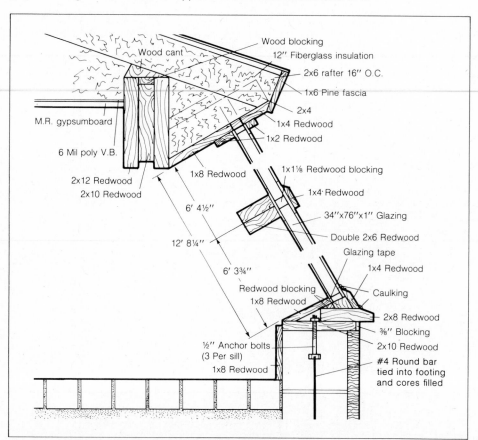

The framing for the glazing is understandably complex. Beveled redwood strips hold the outer panes in position. Doubled 2x6 beams hold the inner panes in place.

passive solar collection and storage systems, and living space all within a greenhouse.

How Much Heat is Gained?

The amount of heating benefits gained by the addition of the greenhouse varies according to the method by which the calculations are made. Mr. Kucko believes the greenhouse will supply between 15 and 25% of the heating requirements of the home. He finds this satisfactory for his installation, for several reasons. Mainly, he finds the heat production adequate when considered in terms of BTU's per square foot of glazing. He knows that to get a higher benefit he would need a much larger glass area, simply because he lives in a very cold climate that requires high amounts of heating. In a new home, he could achieve this amount of glazing; in his situation, however, he feels he has

done the best that is physically possible at a reasonable cost.

THE HEAT STORAGE SYSTEM

The heat storage system is comprised of a floor of brick pavers set in 6 inches of sand. In addition, set along the back (north) wall of the greenhouse are 100 Texxor Heat Cell cans. The glass of the greenhouse consists of six 1-inch tempered, insulated patio door glass panes. These are at a 60° angle to collect the most sunlight during the heating seasons. When the glass of the entryway door is included, there is a total of 128 square feet of collection surface.

CONSTRUCTION DETAILS
Glass Installation

The detailing for the installation of the glass proved to be quite critical. Mr. Kucko had two main concerns. The construction had to be such that any pane

of glass would be easy to remove; at the same time, the construction had to prevent water leakage, which was an extra problem because of the angle of the panes.

The uppermost part of the greenhouse is made of 2x6 joists with 6 inches of fiberglass insulation between joists. In the summer, this solid roof projection shades the storage cells and the windows in the back wall from the full heat of the overhead sun. In winter, however, the roof is designed to allow full solar penetration.

Ducting and Ventilation

In conjunction with wall ventilators in the end walls, a ventilator is located in the roof to provide summer ventilation. Ducting is run beneath the sand bed and in the upper portion of the common house wall. This cycles the cool house air into the greenhouse to be warmed and then routes the air back into the house to supplement other heat sources.

Foundation and Insulation

The greenhouse is set on an 8-inch thick concrete masonry foundation with 2 inches of styrofoam insulation all around. This insulation is covered with fiberbond (see "Insulation" chapter) in all areas that will be exposed to sunlight. The end walls are constructed to match the rest of the house, with 3½ inches of fiberglass insulation in the stud cavities.

RESULTING ENERGY BENEFITS

Some temperature monitoring has taken place since the project was enclosed last fall. It was noted that the lowest temperature within the space reached 22 degrees above zero. This was on one of the colder days and was prior to the installation of both the heat cells along the north wall and the insulating shades on the glass. The warmest temperature was 87° with an outside temperature of +8°. The thermometer was located on the floor and out of the direct sunlight.

A comparison of energy usage for space heating requirements was conducted, comparing March and April of 1981 with March and April of 1980. In doing so, it was found that with the addition of the solar live-in collector, added insulation and weatherizing, 65% less energy was used this year compared to last year for that two-month period.

CONSTRUCTION WORK SEQUENCE

Mid-September 1980	Contract negotiations with DSE
September 30	Excavation and footings formed
October 1	Poured footings
October 3	Laid foundation walls
October 8	Gary put insulation on foundation
October 9	Gary put fiberbond on insulation and installed underground return air duct
October 13	Back filled around foundation—rained that night
October 15	Gary threw in sand for floor
October 16	Sill plates installed—rain is delaying progress
October 17	Framing is started; more rain, more delay
October 18	Large box elder tree downed—was partially rotted and right in front of glass section
October 20	Main beam installed, roof rafters and wall framing in southwest corner erected
October 21	Ceiling joists, roof deck and roofing paper installed
October 22	Shingled roof
October 23	Framed sloping endwalls
October 24	Tilted beams installed
October 27	Framed openings for patio door glass panels
October 28	Installed patio door glass; exterior sheathing applied
October 29	Started exterior redwood steps
October 30	Finished exterior redwood steps
November 18	One interior patio door glass broke (first cold night)
November 30	Gary installed gable louvers in sidewalls
December 1	Exterior walls and ceilings insulated and vapor barrier applied
December 2	Drywall installed on ceiling
December 3-5	Drywall applied to walls
December 10-19	Drywall taped and ceiling sprayed
December 27	Brick floor laid
December 29	Exterior door hung
December 30-31	Wood trim (base doors and windows) installed
January 5, 1981	Interior step built
January 6	Insulated doors for gable louvers installed
April 6	Site inspection by Albright, ED&D Program Manager

7

Window Protection

Windows are one of the largest causes of energy loss in our homes today. Exact figures are hard to pin down, but estimates indicate that windows account for 15 to 35% of the heat loss in homes in the northern United States.

QUALITIES OF A MODERN WINDOW
Glazing
To obtain good performance from a window, it must have natural insulating qualities, built-in weatherstripping and an energy efficient glazing system.

Quality-made windows include double glazing, or two panes of glass with an air space between them. This type of glazing is twice as effective in reducing heat loss as a single pane of glass. The second pane can either be removable, or sealed directly into the sash. In extremely cold climates, triple glazing, which refers to three panes of glass, provides even more insulating value. In warmer climates, tinted or reflective glass reduces glare and heat gain; at the same time the glass permits plenty of daylight. In whatever climate, modern

windows have built-in weatherstripping designed to reduce air infiltration around the sash.

Frame and Sash Materials
There are two basic types of sash and frame material: wood and metal. Wood is a natural insulator and does not require a "thermal break" to keep heat and cold from flowing through the frame. While wood is best as an insulator, it will deteriorate over time and must be painted or stained periodically for protection. Some manufacturers cover the

Windows affect more than just the appearance of a home. Since heat can radiate out as well as in through glass, proper window care is an important part of a heating and cooling system.

wood frames with vinyl on the outside to preserve the wood and simplify maintenance.

Aluminum, the most commonly used window metal, conducts heat 1,770 times faster than wood. The better metal windows incorporate a thermal break, usually made of plastic, which interrupts the conduction of heat or cold through the metal. A metal window without a thermal break will feel cold to the touch in the winter, and will cause moisture to condense on the inside and possibly drip on curtains, walls and carpets and stain them.

In general, the older the house, the more there is to be gained from improved window protection. While everyone cannot replace all their windows with modern new units, there is much that can be done to make the older ones more energy-efficient. Even new ones can be improved.

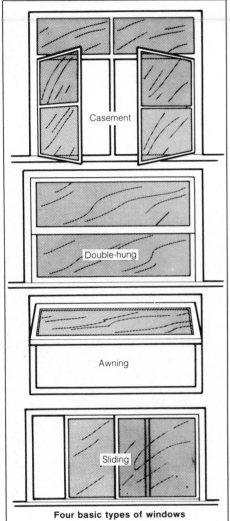

Four basic types of windows

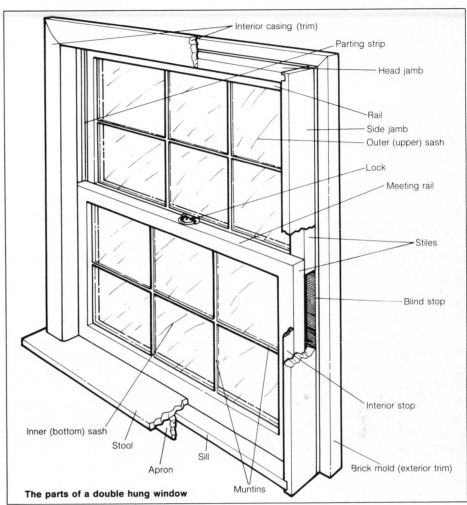

The parts of a double hung window

Shown are the parts of a typical double-hung window.

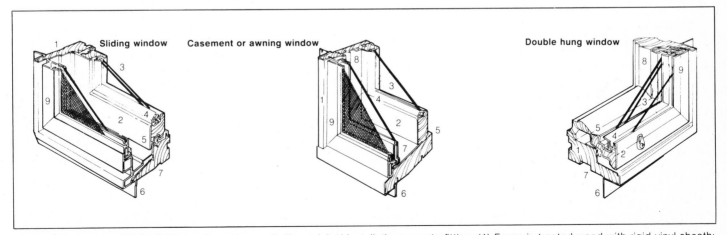

Try to buy windows with the features shown here, whether original installations or retrofitting; (1) Frame is treated wood with rigid vinyl sheath; (2) Wood sash. Outside covered with vinyl; inside natural wood or vinyl covered; (3) Double pane glazing; (4) Vinyl snap in glazing bead; (5) Weatherstripping; (6) Anchoring flange and flashing; (7) Sill; (8) Inside jamb (casement) or stop (double hung); (9) Triple glazing system (storm window). Sliding and double hung have combination screens for insect protection.

SEALING THE WINDOW FOR A VERTICAL FIT

If you have double-hung windows, fasten the sash lock and then try raising and lowering the two sash together. If you can move them up and down even a little bit, then they do not fit tightly at the top and bottom. The result is an air leak. The same is true if the windows move from side to side.

Correcting the Lock Make sure that the sash lock is screwed firmly to each sash. If the screws are loose, remove them; tap in a sliver of wood and replace the screws. If the window still moves up and down a little, you can take out some slack by shimming the sash lock. Remove the lock from the inside lower sash, place a piece of ⅛ inch hardboard or a piece of thick cardboard between the lock and the window frame. Replace the lock. When the lock is latched, the inside half of the lock will raise the outside half of the sash. The resulting pressure will force the inside sash downward enough so that the window will fit tightly at the top and bottom.

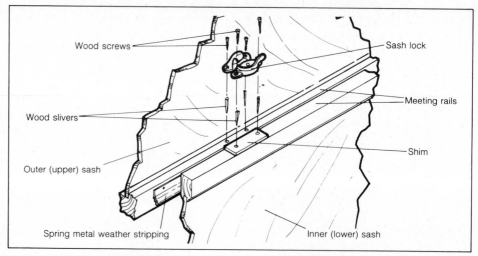

To tighten the sash, reinstall the lock on the rails. A shim will raise the outer sash and push down the lower sash, tightening the two against the jamb and sill.

INSTALLING NEW SASH GUIDES FOR OLD

If the window moves from side to side, the sash guides are worn. If the sash are loose in the channels or guides, but the window sash are good, new channels will correct the problem. The channels are made of metal or vinyl.

Step 1: Removing the Stops First, use a pry bar or butt chisel to carefully remove the stops on the sides and top of the window. You will use these stops again, so avoid damaging them. If they are rotted or too damaged, cut new ones from fresh pine stock.

Pull the windows free. Take out any other stops and clean away old paint. Before installing the new channels, seal any cracks and repaint the area.

Step 2: Removing an Old Cord-and-Weight System If your windows have the old fashioned cord-and-weight balance system, you can remove the devices since the new channels will have their own balance system. Pull the window toward you and until you can see the knot in the cord, which is in a channel near the top of the sash. Near the base of the window frame, you will find a small door. Have a helper open this door and pull the weight and cord out through the opening.

Step 3: Installing the Channels Insert the new channels in the window opening. If they are too long, cut off the bottom with tin snips. The cut must be angled to fit the angle of the sill.

Remove the channels from the opening and fit the sash into them. Remember that the top sash is on the outside and the lower sash is on the inside. Slide the assembly into the window opening. If it goes in easily, drive a tack or small nail at the top and bottom of each channel. Slide the sash up and down to be sure they operate smoothly. If they bind, remove the assembly, and plane one side of the sash. Repeat this operation until the windows operate smoothly, but fit snugly.

Step 4: Finishing the Installation Remove the tacks and replace them with brads, tapping them until the heads are just below the surfaces of the channels. Replace the stops on the inside and outside of the window. They should be snug, but not so tight that the sash bind. If the sash need painting, it is a lot easier to do this before they are seated. Fit the sash so that they work easily, then remove them and paint.

To replace sash guides or channels, remove the stops, install the replacement channel, and replace the inner and outer stops.

Fit the window and channels in the window opening. When they fit properly, fasten the channels with brads, replace stops.

INSTALLING REPLACEMENT WINDOWS

This is the most extensive renovation you can make to your windows, but if your old ones are badly deteriorated, you should consider this option. Many models are available; companies often plan to do the installation themselves. But other manufacturers make windows for do-it-yourself installation.

You will find all-aluminum, vinyl, wood, and combinations of wood and vinyl and vinyl and aluminum. Your best choice is all-wood or vinyl-clad

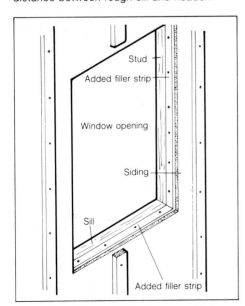

1 A replacement window must fit into the rough opening. Remove interior trim; measure distance between studs and either side and distance between rough sill and header.

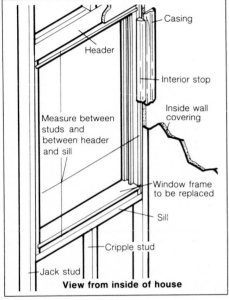

2 Fill the space between the new window and the rough opening. This will not be necessary if the new window is the same size as the old one, but with old houses this is unlikely.

wood. You should only use an all-aluminum window if it incorporates a thermal break, so that outside temperatures are not transferred into the house through the frame. Some windows are made in which the outside of the window is aluminum and the inside is wood. This system also works well. Most replacement windows use double-glazed glass and some have triple glazing; double-glazed glass is worth the cost.

Installing a replacement window takes only a few hours. All models are "pre-hung". They come complete with the sash, jambs, sill and hardware pre-assembled. The windows are available in all the basic styles: double hung, casement, awning and sliding.

If your windows are a standard size, then a replacement window can probably be found that will fit. If not, then the window may require trimming or the frame may require shimming, depending on the fitting provisions made by the manufacturer.

Step 1: Buying Replacement Windows Replacement windows can be of two different types. One consists of new sash guides and new window sash. The installation procedures are the same as for new sash guides discussed above, except that you do not use your old sash, but use the new ones provided. In almost all cases the new sash are double or triple glazed. Many of them have provisions to remove the sash from the inside of the house for ease of cleaning.

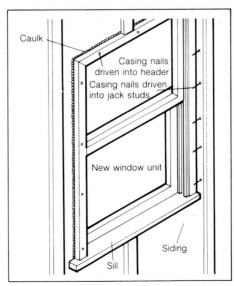

3 Level the new window; fasten it in place with 8d casing nails driven in the top and side trim at 12 inch intervals. Apply caulk around the edges of the brickmold and sill where they join the siding.

This type of window installation offers a good opportunity to remove the inside trim and the apron from under the stool and to insulate the space between the window frame and the studs, header and sill.

The second type of replacement window is designed to replace the entire old window, including the frame. Here you will need to remove all the trim from the inside and outside of the old window, remove the frame and sash and replace it with the new unit. This is a more extensive renovation, but it can be accomplished by a do-it-yourselfer, with the assistance of a helper.

All that is necessary is to select the size and style of the window that most nearly fits the opening left by the removal of the old window. To obtain the measurements for the new window, remove the trim from the inside of the old window and measure the distance between the studs on each side and between the header and the sill. Each manufacturer supplies a variety of filler materials that can be used in the area between the unit and the opening.

Step 2: Preparing the Rough Opening First, remove the old trim and the old window frame. The opening you see, framed with headers and studs, is called the rough opening. Inspect the studs, header and sill for damage from insects or rot. Make any necessary repairs.

Step 3: Preparing the Unit Lay the new unit on a flat surface with the outside of the window facing down. Check the window for square with a steel square. If it is not square, gently tap or push it on opposite corners until it is square. To keep the shape, temporarily attach a 1x2 brace diagonally between the head of the frame and side jamb, and another brace between the side jamb and sill. Use small nails to secure the braces; place the nails so that the holes will be covered by the inside trim.

Step 4: Inserting the Window Measure the outside dimensions of the window jambs of the new window and determine the thickness of the filler material you must add to the rough opening. Use the filler material provided with the window unit, or cut wood filler strips from stock material.

Insert the window unit into the opening from the outside. With a helper, push the window unit upward until it fits tight against the header and is centered between the studs or filler strips.

With a level, find which of the top two corners is lowest. Drive a 8d casing nail through the brick mold (the outside trim) into the header to anchor this corner. Level the window and anchor the other top corner.

On the inside, shim the space between the rough sill and finish sill and fasten the corners with 8d casing nails. Also shim the sides of the window in at least two places on each side. Work carefully. Do not insert shims thicker than the space between the opening and the frame, or the window frame may bow and cause the sash to bind.

On the outside, nail the brick mold to the rough frame with 8d casing nails every 10 inches or so. Countersink the nails and fill the holes with wood putty.

Vinyl covered frames Windows made with a vinyl covered exterior wood frame have other provisions for attaching them to the rough opening so the vinyl is not punctured with a nail. Follow the directions provided with the unit.

Step 5: Sealing and Shimming If the brick mold on the new window does not reach to the house siding on each side of the window, fill in the space there with wood shims. Caulk the spaces and

cracks between the shims, siding and brickmold.

If your walls are thicker than those found in today's construction, the jambs may not reach the face of the inside wall. In this case, nail strips of wood called jamb extensions to the inner edges of the jambs. If you cannot obtain jamb extensions, cut your own from ¾-inch wood stock. Fasten the jamb extension, sawed side out. Finish the rough edges with sandpaper.

Now fill the space between the jambs and the rough opening with insulation pushed in place with a putty knife. Fill every space, no matter how small. Cover the space between the jamb and rough frame with a strip of 4-mil plastic to provide a vapor barrier. Fasten the barrier in place with staples.

Step 6: Finishing Off Finally, replace the inside trim; then paint, stain or finish the window to match your decor.

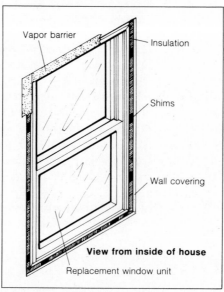

View from inside of house

Vapor barrier · Insulation · Shims · Wall covering · Replacement window unit

4 Fill the space between the jambs and rough opening with insulation. Then staple a strip of plastic vapor barrier over the insulation. The final step is to replace the trim.

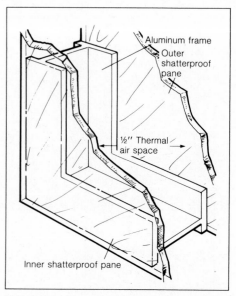

Aluminum frame · Outer shatterproof pane · ½″ Thermal air space · Inner shatterproof pane

5 A double glazed window is much more effective an insulator than a single glazed one. The two panes are separated by a ½ inch air space, which blocks heat transmission.

SEALING DOUBLE HUNG WINDOWS FOR DRAFTS It is not always necessary or practical to replace windows, especially when all that is needed is weatherstripping. There are several types of weatherstripping that can be applied to a double hung window without special tools: vinyl in tubular or foam filled strips, adhesive-backed foam rubber, thin spring metal strips and various type of materials held by wood and metal strips.

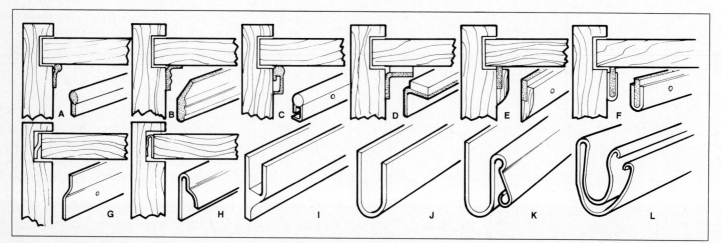

Various types of weatherstripping. A, Tubular or foam filled vinyl strip. B, Molded vinyl strip. C, Vinyl tube on metal strip. D, Fabric on metal strip. E, Felt or foam recessing in wood molding. F, Felt held in aluminum strip (comes in roll). G and H, Two types of spring metal for double hung windows. I and J, Vinyl weatherstripping for casement windows. K and L, Spring metal for casement windows.

Installing Vinyl Strips The "tube" portion of the vinyl strip can either be hollow or filled with foam. The flange portion of the strip is tacked or stapled to the parting strips on the outside of the sash and along their sides. It is attached to the sash themselves at the top, bottom and center.

Installing Adhesive-backed Foam Adhesive-backed foam can be attached to the bottom of the lower sash, and at the top of the upper sash for windows that do not fit tightly at the top and bottom. However, this type of weatherstripping does not do an effective job of sealing the sides.

Installing Spring Metal Strips Spring metal strips are placed in the sash channels. The window sash need not be removed to install the strips. Nail the metal strips to the channels so the spring opens towards the outside. Do not cover pulleys at the top of the channels; trim to fit around them.

Nail a strip the full length of the sash across the bottom edge of the lower sash and across the top of the upper sash. Attach another strip to the inside edge of the lower rail of the upper sash, with the spring facing down. This seals the space between the sash where they join in the middle.

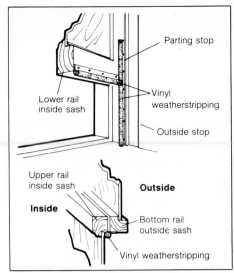

Tubular vinyl strips are attached to the top, center and bottom of sash and to the parting stops along the sides.

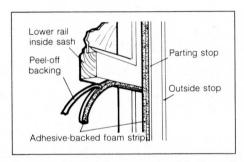

Adhesive backed foam weatherstripping will seal the top and bottom of windows when applied to their top and bottom rails.

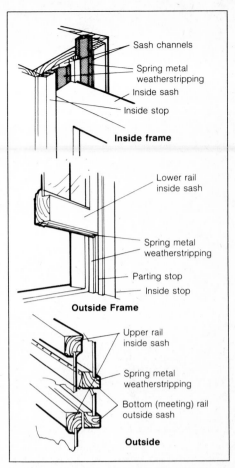

Spring metal weatherstripping fits in the sash channels, between the meeting rails and along the upper and lower rails. Spring metal stops infiltration but transmits cold.

SEALING OTHER TYPES OF WINDOWS

Casement and awning windows cannot normally be sealed by placing weatherstripping on the outside. In most cases, these windows can be made weathertight by adjusting the locking handle. If that does not seal the window, place adhesive-backed foam weatherstripping between the sash and the frame. Because of the thickness of the foam, you may have to readjust the locking handle to allow for this thickness. Otherwise the handle will not latch.

Horizontal sliding windows usually have a fixed sash and an operating sash. Caulk the fixed sash from the outside. Then, seal the operating sash by nailing tubular vinyl strips along the top and bottom frames, the leading edge of the sliding sash and the front edge of the fixed sash.

Jalousie windows are difficult to seal; the only effective weatherstripping is storm windows. A transparent vinyl weatherstripping is available which helps to a limited degree. It is placed on the edges of the glass louvers and fits between the panes when they are closed. Because jalousie windows are so difficult to seal, they should only be used in warm climates or in rooms or porches that you do not need to heat or cool.

STORM WINDOWS

One of the most significant ways to reduce air infiltration around windows is to add storm windows. They also increase the R value by adding another pane of glass and another air space. The air space between a window and a storm window should not be any wider than two inches, or the space begins to lose its insulating effectiveness due to convection currents set up between the glass panes. A tight-fitting storm window can reduce infiltration around windows by as much as 50%.

We have typically thought of storm windows as being placed on the outside of double hung windows. Storms were made of wood or aluminum frames and used glass or plastic for glazing. Today most ready-made "storms" are made of aluminum if they are to be mounted on the outside of the window. Wooden storms are difficult to find ready-made, and custom made ones are very expensive, but for the person restoring the old house, they can be found in kit form. Plastic storm windows of many different designs are now made, and aluminum-framed storms are also available for inside mounting.

MAKING ALUMINUM STORM WINDOWS

Aluminum storm sash can be made at home and the cost will be much less than that for a readymade storm. The sash are available in hardware and home center stores along with the glazing channel that seals the glass in the sash. Single-weight glass is satisfactory for windows that are less than 9 sq. ft. in size, although double weight glass is better and is recommended. If the window exceeds 9 sq. ft. in size, it should be made as a two-piece sash, by adding a divider in the middle. Extra tall windows are best built in two or more sections for added strength.

Tools and Materials You will need the following tools to make aluminum storm windows: A hack saw with a fine-tooth blade (32 tooth per inch), file, combination square, sharp razor blade or utility knife, screwdriver and a rubber or wooden mallet. If you are making several sash, a miter box for use with the hack saw would be convenient. Glass must be cut accurately, usually $1\frac{1}{16}$ inches less than the outside measurements of the sash, but this may vary with different manufacturers of the aluminum sash.

Step 1: Finding Dimensions The measurements and positions of the storm windows will differ according to the type of window.

Double hung windows The measurements should reach just outside the window stop so the windows will fit against the stop when they are hung. Subtract $\frac{1}{8}$ inch from these measurements to allow for cutting errors. For storms that are made of more than one vertical section, allow $\frac{1}{8}$ inch clearance in height for each section.

Casement windows These are fitted with storms on the inside. Usually the casement has holes along the sides and in the middle. These holes hold clips to keep the storm sash in place. Allow an extra $\frac{1}{4}$ inch height and width for frames for these windows.

Awnings Storm sash for awning windows also are attached on the inside and are screwed to the window stop. Measure the distance between the stops and from the crank to the top. Allow $\frac{1}{8}$-inch clearance in height and width.

Some awning windows do not have a stop at the bottom of the window. Then you must add a wood filler strip. Cut the

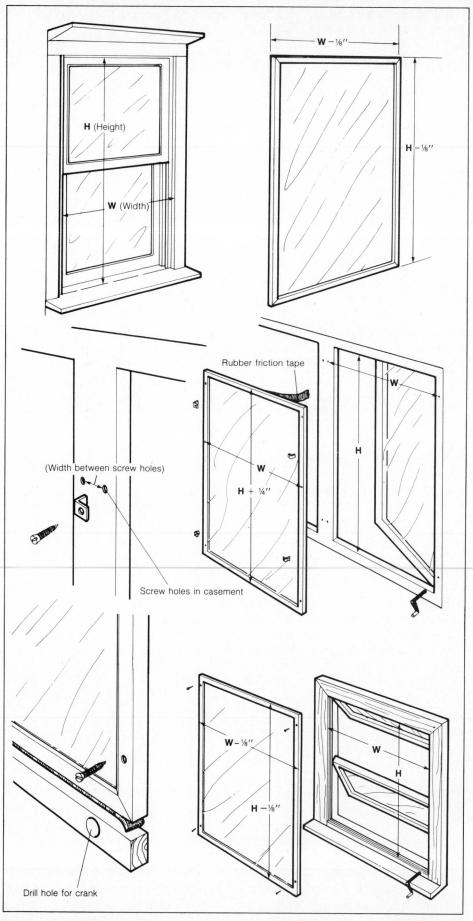

Measure windows and allow proper amount of clearance. Long windows should be fitted with center brace.

filler strip equal to the width of the window frame. Cut a hole in the filler strip for the opening crank, and screw the strip to the window frame. The storm sash is ⅛ inch narrower than the window and ¼ inch less than the distance from the filler strip to the top of the window opening. Self-adhesive weatherstripping is added to the bottom of the storm sash to fit against the top of the filler strip.

Step 2: Cutting the Sash The sash will come with a plastic glazing channel to seal the glass in the aluminum channel. Remove the glazing before cutting the sash. Use a combination square to mark the sash at each corner for a 45 degree angle. Cut on the scribed lines with the hack saw; then file off the burrs.

Step 3: Assembling the Sash The sash are held together with friction fit corner locks. Insert locks in both corners of the short sides of the sash. Fit the plastic glazing around the glass, holding the glazing temporarily in position with tape. Make 45° cuts at the corners of the glazing with a sharp knife, but cut only the edges to produce a mitered joint. Fit the sash over the edges of the glass, pressing the sides together until the corner locks are in place. It may be necessary to use a mallet to get a tight fit. Do not use a regular hammer, or the frame may be damaged.

Step 4: Weatherstripping the Sash Place adhesive weatherstripping around the inside edges of the storm to fit against the window stop. Without the weatherstripping, no matter how accurate your construction, the storm will not have a good seal.

Step 5: Hanging Storm Windows Hang the storm window using the brackets made for them. Spring clips hold the storm window from the inside. The clips hook over special screws driven into the sill. Long storm windows will seal better in the center if turn buttons are used to hold the storm sash tightly against the window stop. The turn button screws to the window frame.

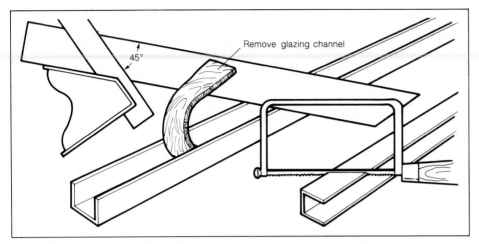

Mark a 45 degree angle at the appropriate points. Then, using a fine-toothed blade in a coping saw or a hacksaw, cut along the mark.

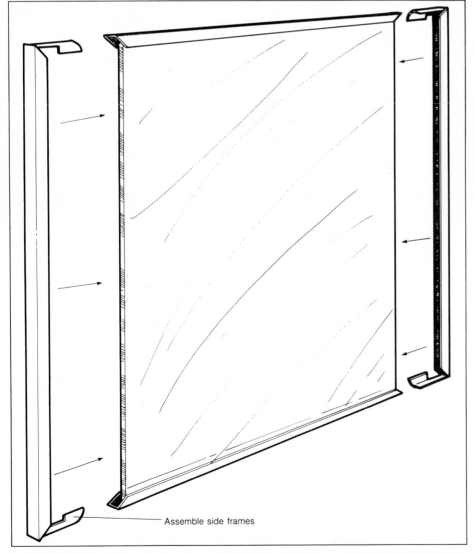

The glass is installed in the channel. Assembling the glass and channel is easier if it is lubricated with a 50-50 mixture of water and dish washing detergent.

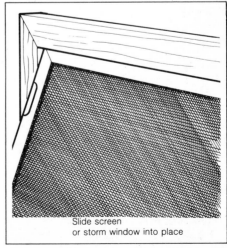

Hang the window using the bracket provided. Supplement with self adhesive foam strip.

HOW TO BUILD A WOODEN STORM WINDOW

For double hung windows, storm sash made of wood are even better than aluminum ones. Wood storms are more difficult to make, and they require painting from time to time, but many homes still use them.

Tools and Materials To make the wooden storm, you will need dowels or screws, ⅜-inch drill, waterproof glue, pipe clamps, router, wood chisel, hammer, latex caulk and quarter round, as well as glass and the wood for the frame. The best wood to use for the frame is 5/4 (1⅛ inch thick) clear pine. It is straight grained and does not have any knots. The glass will be cut ⅛ inch less than the distance between rabbets cut into the assembled frame.

Step 1: Finding Dimensions The sash width should be the same width or slightly narrower than the window sash for best appearance. In most cases, this will be 1½ or 2 inches wide. The wooden sash fit against the window stop and are measured in the same way as the aluminum sash. Allow ¼ inch clearance for construction errors and possible swelling of the wood when it absorbs moisture.

Step 2: Preparing the Members Cut the vertical stiles full length. The rails and muntin (interior crosspieces) fit inside the stiles. Select the method of joinery that you prefer. Pegged mortise and tenon is best, but joints that are secured with dowels or screws are easier to make. Metal mending plates can be used to reinforce the corners, if desired.

Step 3: Assembling the Frames Position the rails and stiles. If you are making several windows, you can make a support to assist in holding the frames. Drill holes through the stiles and into the rails and muntin to a depth of 1½ inches for ⅜-inch dowels. You will have two dowels in each joint. Cover the dowels with waterproof glue; install the dowels and clamp the pieces with pipe clamps. After the glue has set, use a router with a ⅜-inch rabbet bit to cut rabbets ½ inch deep for the glass. You will have to finish the corners with a wood chisel and hammer. Prime the window before installing the glass. This prevents moisture infiltration.

Step 4: Installing the Glass After the primer has dried, apply a thin layer of latex caulk in the rabbets; put the glass in place. Press it down until the caulk seals all the edges. Apply another thin layer of caulk along the edge between the glass and the frame. Cut ⅜ inch quarter-round with mitered corners to fit the rabbets and nail them in place with ¾-inch brads. Clean up any excess caulk; then apply the final coat of paint.

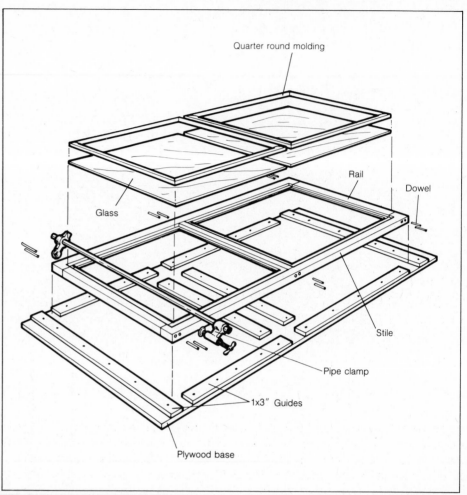

Wooden storm windows are difficult to find, and are expensive. They are less convenient than aluminum ready-made storms, but will seal better and cost less if you make your own.

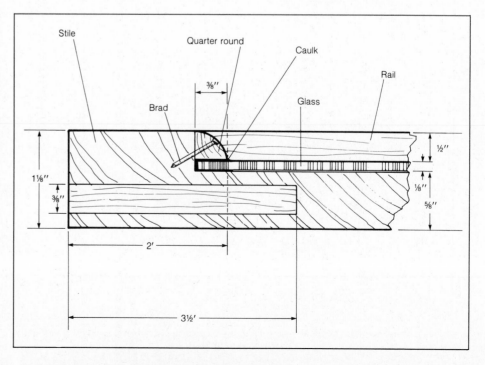

BUYING ALUMINUM STORM SASH Aluminum combination screen and storm windows are available ready-made to be installed on the outside of double hung windows. The storms consist of two glass or plastic panels and one screen panel. One or more of the glass panels slide, depending on the type of frame. Double track frames have a glass panel and screen panel in the outside track, the glass at the top and the screen at the bottom. The second glass panel is on the inside track and it is the only one that moves. A triple-track storm has separate tracks for each panel and usually allows any of the panels to be moved independently of the other two.

Purchasing the correct size combination storm sash is important for proper fit. If you already have screens or old storm sash on your windows, remove them and measure the frames. If there are no storms, measure the height and width between the window frames, just inside the window stop. Aluminum-framed storm sash are made so they can be trimmed slightly around the edges with tin snips, to fit windows that are slightly off-size or have minor irregularities. This is important for an older home.

Ready-made aluminum storm windows are installed on the outside with screws along the sides and top. Note two drain holes at bottom of each storm just above sill.

INSTALLING READY-MADE STORMS To position a ready-made home storm sash, remove the panels as for cleaning. Trim the frame to fit the window frame opening. Apply a bead of caulk along the window stop, across the top and down the sides. Place the storm sash frame in position so that the caulk seals the space between the window stop and storm frame. To hold the frame, put one wood screw through one of the prepunched holes in the top center of the frame. Then drive the rest of the screws along the top and sides. Clean the window panels and insert them and the screen panel in the frame from the inside of the house.

There usually is no provision to seal the bottom of the storm sash where it meets the window sill. In fact, most storm sash have a small indentation, or drain port, that the manufacturer does not want you to seal. This indentation permits rain water to drain from the bottom of the window frame area when one of the window panels is open. In the winter, when the storm sash are closed, place a bead of caulk along the bottom of the sash to seal it. If the storm sash will be opened in the summer for ventilation, remove the caulk at the drain port. Storm sash that will not be opened in the summer need not have the caulk removed.

INTERIOR INSULATION PANELS

Numerous devices have been developed so you can insulate windows from the inside of the house. They vary from aluminum storm sash (similar to outside storm sash) that fit in the inside of the window frames, self-adhesive plastic frames holding clear plastic panels that stick to the window frame, to insulating window shades and shutters.

A new energy-efficient storm window for outside installation has an expander framing system, so it is adaptable to older, out-of-square windows.

One of the easiest ways to insulate windows is with inside storm sash. This one uses aluminum frame with clear plastic insulating film.

BUILDING ALUMINUM FRAME PANELS

Measure the dimensions of the window. Subtract ¼ inch from the overall height and width. This subtraction allows enough space for foam tape insulation that goes around the frame. Construct the frame as for the exterior aluminum storm windows (above). Apply ¼x⅜ inch self-adhesive foam tape completely around the frame edge.

Place the frame into the window frame from the inside. It should fit snugly, with the foam tape creating a tight seal. If the window is difficult to seat, use a wide blade putty knife or a kitchen spatula to squeeze the foam tape in place.

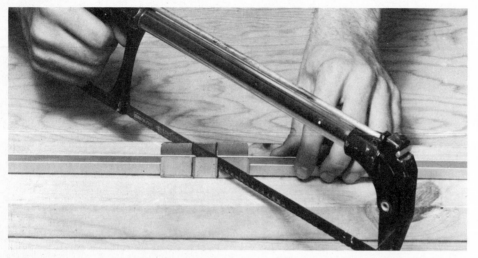

Using a mitre box and hacksaw, cut the aluminum sections to correspond to the window measurements. Remove splines before cutting.

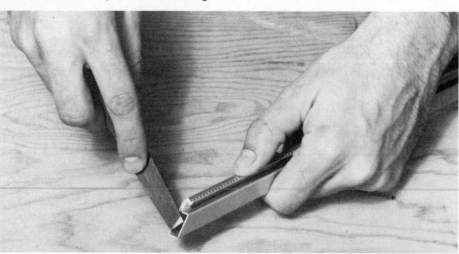

Debur the rough edges with a fine-tooth file to ease installation of the corner locks.

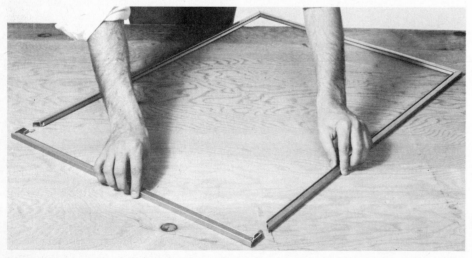

Insert corner locks in short lengths; then attach to long lengths to form a square. Use rubber mallet to tap in locks, if needed.

Cut insulating film one inch bigger all around than the completed frame. Place frame groove side up, and lay the film across the frame.

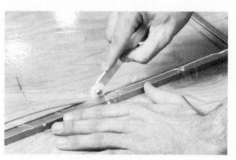

Starting in one corner, lightly press the spline into the groove. Repeat this on the opposite side, keeping the film taut. Repeat the procedure on the other two sides.

Use a utility knife to trim off the excess close to the outside edge of the spline. Take care not to pierce the insulating film.

Add the foam strips to the panels. Peel off the paper backing. Press the sticky side of the tape to the outer edges of the frame.

MAKING INTERIOR ALUMINUM FRAMES

You also can build aluminum frames that hold a flexible plastic film held in the frame by a plastic spline. These insulating panels are simple to make. A single panel can be assembled in less than 15 minutes. The kits provide all the necessary materials, or you can buy the materials individually. An identical panel using fiberglass solar screens can be made to hang outside to cut sun infiltration and reduce air conditioning costs.

Tools and Materials You will need the following materials and tools: aluminum sections with spline (available in 6 and 8-foot sections), corner locks (4 per panel), insulating film (available in 36- or 48-inch widths), center brace sections for each window more than 3 feet long, ¼-inch thick x ⅜-inch wide self adhesive foam tape, miter box and spline insertion wheel, rubber mallet, hacksaw with fine blade, metal cutting file, utility knife or single edge razor blade, and a tape measure.

Step 1: Constructing the Frame Measure your windows as shown. Remove the spline from the sections before cutting. Using the miter box, cut the aluminum sections to correspond to the measurements you have made. Smooth rough edges on the frame with a fine-tooth metal file. This makes it easier to insert the corner locks. Insert the corner locks in the short aluminum sections; connect them with the long sections to make a square. If necessary, use a rubber mallet to tap in the corner locks. Do not bend the aluminum sections.

Inserting a center brace If the window is more than three feet long, you should install a center supporting brace. Cut a center brace to the inside width of the frame and snap the brace in place at the center point of the side frames. The brace must be snapped in place before inserting the film.

Step 2: Inserting the Plastic Film Cut a section of film one inch bigger all around than the completed frame. Place the frame groove side up and place the film over the frame. Beginning in one corner, lightly press the spline in the groove. Proceed to the next corner on the same side. Use a spline roller to roll the spline into the frame till it locks in place. Repeat this process on the opposite side. Keep the film taut as you insert the spline, but no so tight that the frame is bowed. Repeat this process on the remaining two sides. Carefully trim the film next to the outside edge of the spline.

The manufacturer of the "In-sider" storm window claims it will save 98 percent of the energy dollars wasted by uninsulated windows. Self adhesive plastic frames snap closed to hold plastic panel in place.

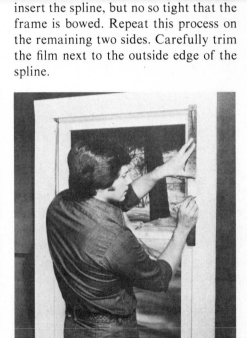

Measure the window from the top of the sill to a top point flat enough to accept the molding. Measure the distance between side casings.

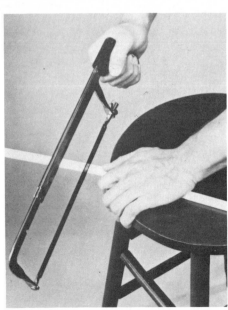

Cut the frame and sill moldings to the proper length. The top molding and the sill molding fit inside and flush with the side moldings.

Mark the plastic panel to size; score with sharp knife. Snap the panel along the scored line.

Press the adhesive molding to the casing; position the plastic sheet; snap the frames closed.

Step 3: Applying the Foam Tape

Apply foam tape around the edges of the frame. Remove the protective covering on the tape; then, starting in one corner, stick the tape to the outside edge of the frame. For a proper fit, avoid stretching the tape. The frames are now ready to install in the windows, just for the other aluminum-framed panels.

Aluminum frame interior storm window can be hinged open either from side or from the top.

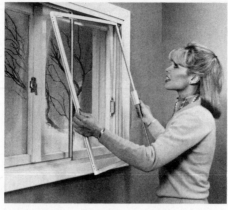

Sliding version of Aluminum In-sider is for sliding windows. It uses glass or plastic glazing.

This interior storm window uses film stored on a roller. The film, on vertical runners, can be lowered or raised like a window shade.

BUILDING PLASTIC FRAMED PANELS

Typically, a self-adhesive molding (frame) that holds a semi-rigid clear plastic sheet is applied to the window frame and sill. The plastic frames are self-hinging so they can be opened for removal or replacement of the plastic sheet. The panels completely seal the window to stop infiltration and create an additional insulating air space.

Step 1: Making the Frame Measure the height of the window from the top edge of the frame to the top of the sill and the width from the outer edges of the sides of the frame. The plastic needs a flat surface to which to adhere, and this is usually on the outside of the window frame.

Step 2: Cutting the Plastic Cut the plastic frame strips to size with a sharp knife or a fine-bladed hacksaw. The frame molding corners can be square or mitered.

Measure the plastic panel to size. A grease pencil or crayon works best for marking on the plastic. Thin plastic sheet can be cut with a sharp knife or scissors. To cut the thicker semi-rigid sheets, score along the marked lines with a sharp knife. Then place the sheet on a table with the scored line next to the edge and, while holding the larger part of the sheet firmly to the table, snap off the other part. The sheet will break on the scored line.

Step 3: Installing the Panels You will have to remove any curtain fixtures that attach to the window frame. Replace them either beside or above the window frame. They cannot be put back on top of the plastic molding. Clean the contact area at which the molding will adhere. Place the moldings on the plastic window, making sure the moldings fit properly in the corners. Remove the protective covering from the self-adhesive strip on the moldings and placing the sill molding down first; stick the panel to the window. Then press the frame molding firmly in place.

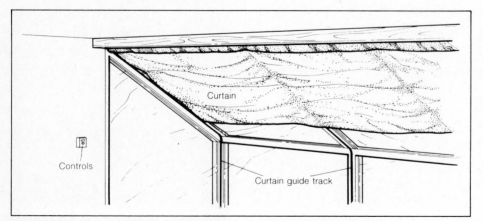

Sun Quilt®, having an R-factor of 5.9, is 1½ inches thick. The three layers consist of an outer layer of beige cotton/polyester blend fabric, a center of hollow-fill fiber insulating material and an inside fabric of metalized fabric with a urethane vapor barrier. The quilt rides on tracks and folds in accordian pleats when retracted. Each 16 inches of vertical height requires 2 inches of storage space. The quilt is operated by a small motor, either manually, or by heat or light sensors. The power for the motor can be provided by solar cells.

USING SHADES, SHUTTERS, BLINDS AND SCREENS

Even if your house has been equipped with storm windows and weather stripping, windows will lose heat to the outside air during a cold winter. Passively heated homes, with their large window area, have a much greater potential loss. One solution is thermal shutters and shades for use at night. They can reduce air infiltration, add additional air space and reduce heat loss through radiation. Some shutters and shades retain heat in the winter and reflect solar radiation in the summer.

There are hundreds of ideas and products on the market for windows, and it would be impossible to discuss each in detail. Many that are designed for do-it-yourself installation and come with installation instructions. Obtain all the details you can on a product before you purchase it. Be sure it will fit your window and that it is designed to accomplish your objective. Remember that since little sunlight comes through a north-facing window in winter, a solar screen, designed to hold the warmth in a solar home will accomplish little on that window. That same screen placed on the west side of the house for use in the summer could be very practical.

Improving the Insulating Effectiveness of Window Shades

Ordinary window shades can be made more effective by installing a small cornice or wood strip across the top of the window, and mounting vertical strips to the window frame. Mount the shade so that when it is rolled down, its surface is toward the wall and against the strips and cornice. The shade does not have to be sealed, but it must be tight enough to reduce air flow in and around it. The shade can stop at the window seal, or, for a decorative effect, run the vertical strips and the shade down to the floor. The bottom of the shade must fit tightly against the stool or floor to be effective.

Window Quilt™ can be used on any size window and retracts on roller at top of window. The quilt is made of five layers: two covers of a heat reflector of a thin sheet of aluminized polyester plastic and two layers of fiberfill insulation. The quilt rides in tracks that are attached to the casing with double faced foam tape. A rigid bar with weatherstrip attached seals at the bottom.

The "Insulating Curtain Wall®" combines thermal insulation with large greenhouse type windows. The assembly is mounted inside above vertical windows. An electric motor opens the curtain automatically on a sunny day and closes the insulating curtain automatically during cloudy, cold daytime periods, and at night. In summer the operation can be reversed. The curtain can provide R-values between 9 and 12.

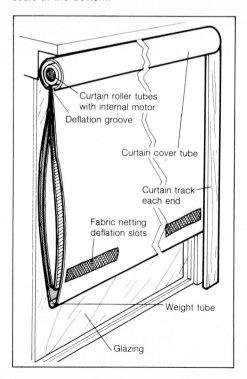

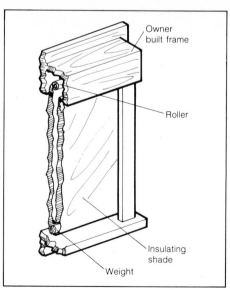

"Independence 10"® is an insulating window system that operates much like a regular roll down shade. The curtain is tightly fitted within a frame that is permanently mounted to the sides and top of the window, creating an effective thermal seal.

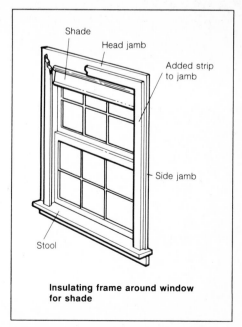

Insulating frame around window for shade

Adding strips of wood to the top and sides of the window jambs reduces air circulation around a window shade.

Window shades and drapes can be effective insulators. Use cornices and frames; fit the shades tight against the stool to cut air flow.

The cornice and vertical strips can be natural wood or covered to match or contrast with the shades.

Improving Heat Retention of Draperies

Many houses use draperies to achieve the same privacy as is offered by shades. Draperies retain heat better than roller shades, but they do have gaps around the sides, top and bottom. This permits air to flow through, reducing their ability to insulate.

The effectiveness of draperies can be improved in ways similar to those described for shades. A cornice at the top, fitting tight against the wall and close to the top edge of the drapery, seals the top. The bottom can be sealed by letting the draperies fit next to the stool and hang to the floor. The sides should fit next to the wall. You can achieve a framed window effect by building a vertical section with the same profile as the cornice.

Draperies themselves hold even more heat if you add liners that are stuffed with fiberglass insulation. But these draperies may be extremely bulky when drawn back to admit light.

Shutters

One of the most effective means of saving energy through window insulation is to add shutters. They can reduce heat loss through single-glazed windows by a factor of six. Even with insulating glass or storm sash, heat loss still is reduced by two-thirds.

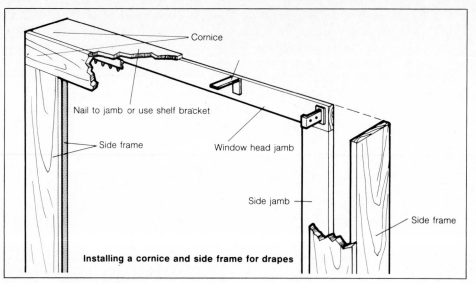

Installing a cornice and side frame for drapes

Adding a cornice and a side frame helps draperies seal a window. Use right-angle metal brackets or nails to support the frame.

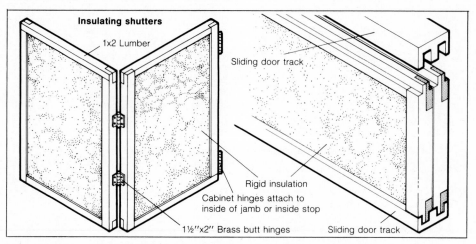

Insulating shutters

Make simple insulating shutters from 1x2 lumber and polystyrene insulation. The shutters can be hinged from the side or from the top, or they can be made to slide.

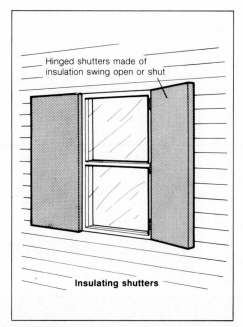

Insulating shutters

Create simple, effective exterior shutters from insulation board. Hinges hold the shutters in place; magnetic locks hold them closed.

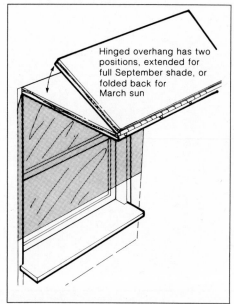

When folded back, a double-hinged overhanging shutter provides the appropriate amount of shade for winter months. When laid open, the shutter keeps out bright summer sun.

MAKING INSULATING SHUTTERS

The shutters can slide, fold sideways or up or down, depending on the location and window size. Size them to fit against the face of the window trim or the inside jamb.

Step 1: Building the Frames The design shown uses 1x2s. You can use any straight-grained wood, from pine to redwood, depending on how you desire to finish the frames. Lap the corners; nail and glue them together.

Step 2: Inserting the Panel The insulation used is ¾-inch closed-cell polyurethane sold in 2x8 feet sheets as building sheathing. Using a sabre saw, cut the insulation so it is about ⅛ inch larger than the opening. In this way, the panel will fit snugly in the frames. You can cover the insulation with a fabric or leave the panel plain. If you glue fabric to the insulation, use white glue diluted 50-50 with water. Or you can staple the fabric to the edges of the insulation. Then press the panel into the frames. The "friction fit" should hold them in place.

Special finishes Other decorating schemes will work well, such as covering the whole frame with decorative paneling to match paneling on the walls. For maximum effectiveness of a shutter that faces out on east or west windows, consider covering the insulation with a reflective material.

Step 3: Connecting the Panels Join the panels with 1½x2-inch brass butt hinges. Two should be used for shutters up to 4 feet long, three for larger shutters. Use cabinet hinges to attach the shutters to the window frames.

Sliding Shutters Sliding shutters are not hinged, but use an inexpensive fiber track and shoe. The top and bottom of the shutter is routed out to accept the shoe, and a track is attached to the sill and to a 1x2 frame attached to the face of the top trim. Sliding door tracks also will work. Use a router to cut a groove along the length of the shutters. This groove functions as a "track" that fits in the channel of the door track.

Solar Screens

Solar screens are available in several different designs. One is placed in an aluminum frame (refer to aluminum-framed panels) and hung on the outside of the window in the manner of a window screen, which purpose it also serves. Others are designed to be used on the inside, operating much like venetian blinds. These screens reduce visual perception, but do not block it completely. They can, by reducing the amount of hot sun that shines through the window, reduce air conditioning costs and damage and fading of decorating materials in the home.

Venetian Blinds

Also popular and effective are new designs of venetian blinds. A few window manufacturers incorporate them into their window designs, sometimes installing the louvres between the panes. Some new designs are mounted to provide a reasonably air tight fit along the edges of the blind to create an air space between the blind and the window. There is a vertical venetian blind, with slats about 3½ inches wide. This model has a reflective coating on one side and decorative fabric on the other. The louvres rotate 180° to control light and to eliminate glare and excessive heat.

BASEMENT WINDOWS

Basement windows are often overlooked. A typical builder's basement window merely snaps into a metal frame with spring clips. The metal frame is set into the poured concrete or cement block foundation; all that can be done to insulate the frame is to caulk around it to prevent air infiltration.

You can use storm windows with these windows, either on the outside or the inside. Any of the different types described in this chapter will fit, but one of the easiest is the plastic panel that is framed with self-hinging plastic molding. The molding attaches directly to the metal frame. The plastic insert can be removed for cleaning or maintenance, and all work on the window can be done from inside the basement.

Sliding insulation panel

Sliding shutters of insulation board move along a track and shoe. Do not use styrofoam or polystyrene for interior shutters, since both are flammable.

Fiberglass solar screening reflects the sun's rays. The screens fit in the same installation as the outside storm window.

Verosol® is a shade that is made of a one-piece, pleated, metallized fabric. It is made of semi-transparent polyester, bonded on one side to a micro thin layer of aluminum. The shade reflects solar heat and acts as an insulator in winter and summer.

8

Installing Insulation

Insulation is one of the few things that will pay for itself. In many cases the "payback" period will be as short as one or two years. Of course, the insulation must be installed properly; it must have an efficient vapor barrier; and, there must be ventilation.

Although insulation resists the passage of heat, it does not resist the passage of moisture, which is the reason you need a vapor barrier. Also, if heat builds up constantly on one side of the insulation, as happens in an attic, the insulation eventually absorbs the heat. Later the insulation will reradiate this heat to the cooler side. In the case of the attic, the cooler side is the house below, especially when the home is air conditioned. The answer to this problem is ventilation, which removes heat before it saturates the insulation.

TYPES OF INSULATION
There are several kinds of insulation, almost all of which can be installed by the homeowner. There is loose fill insulation that can be of fiberglass, mineral wool, cellulose or vermiculite. There also is foam insulation, as well as several new and experimental types. With the exception of foam, the homeowner can install all insulations.

Batts or Blankets
These are sized to fit between studs, joists and rafters that are 24 inches or 16 inches on center. Be sure to purchase the right width. Most roll or batt insulation will have a vapor barrier on one face, with flanges at each side. The flanges staple between the studs, joists or rafters. The vapor barrier always faces into the room.

An alternative design is meant for use under a floor, such as over a crawl space. The flanges will be on the "bottom" face, the one having no vapor barrier. This is so that the vapor barrier can face up toward the floor while the flanges are down, so you can staple them to the wood members.

Mineral Wool
Mineral wool insulation comes in both batt and roll form. The efficiency of mineral wool insulation is similar to fiberglass. When working with either insulation, wear a respirator. The fine particles that break loose from the insulation can be inhaled and be a health hazard.

Cellulose
Cellulose insulation can be blown into walls or on attic floors between joists, or can be poured from a bag. This type of insulation is made by processing used newspapers. The resulting material is treated to make it fire- and vermin-resistant. Cellulose insulation has a higher "R-value" than any insulation except some of the closed cell rigid plastic insulation boards. It also is one of the least expensive insulations.

Despite its low cost and high efficiency, cellulose insulation has an undeserved reputation as a material of low quality that is flammable and supports vermin. The reason for this is that in the past the quality of cellulose insulation was low. It was not properly treated. To be sure any cellulose insulation you buy is properly manufactured, look on the bag for a government specification number HH-1-515C, which indicates it has been manufactured to be fire- and vermin-resistant.

Rigid Board
One of the very newest types of insulation is a rigid slab or board that comes in thicknesses of ½ inch and more. Rigid insulation is used under poured concrete slab floors and around the outside edges

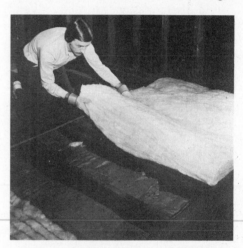

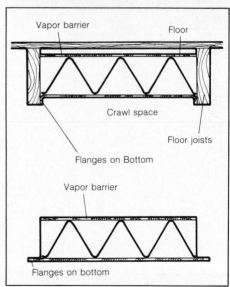

Special insulation variety has flange on the side opposite the vapor barrier. This permits stapling to floor joists with vapor barrier up against the floor.

of these floors. It is used to insulate the inside and outside of basement walls because it has a high R-factor and also is relatively resistant to damage. When it is used on the outside of a basement

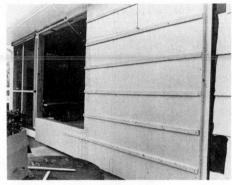

Sheets of rigid insulation can be used as insulating sheathing. However, it is not a nailing surface, so furring strips must be nailed through it into the studs. Space between siding and sheathing is dead air. This adds a bit of insulation factor.

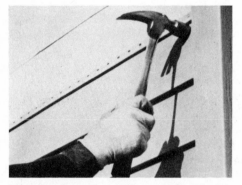

Cover the insulation with new siding. In this situation a good vapor barrier on the inside of exterior walls is absolutely essential.

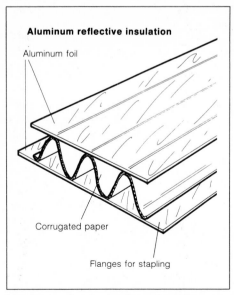

Aluminum reflective insulation

Aluminum foil

Corrugated paper

Flanges for stapling

Aluminum reflective insulation is designed to create dead air space between the two foil sheets. The foil is supposed to reflect heat back into the room; the dead air is supposed to prevent heat loss due to convection.

wall, the board should extend down to the footing, although it is reasonably efficient when placed at least below the frost line.

Underground Basement Walls As noted in earlier chapters, rigid insulation makes any masonry wall a thermal storage unit that can hold thousands of BTUs of heat energy. This heat is released when the surrounding air drops in temperature below that of the wall. If the rigid insulation can be placed outside the foundation, the concrete or concrete block walls of the basement then become "heat sinks" for storing heat energy. Protect any portion of the insulation that is above the ground with sheets of cement asbestos siding or with panels of simulated masonry formulated for exterior use.

Advantages and Disadvantages Rigid insulation commonly is called "pinkboard" or "blueboard" because the major manufacturers of the product color the plastic when they add ultraviolet inhibitors. The inhibitors reduce damage from the sun when the panels are exposed to it.

Another type of rigid insulation, "beadboard," is not colored and quickly deteriorates in the presence of sunlight. Avoid beadboard; it not only is damaged by sunlight, it also absorbs moisture. This reduces its insulating efficiency. Beadboard is white and has a "pebbled" surface.

Use foam insulation for small spaces in which other types of insulation would be difficult to install. The material sets quickly and can be trimmed with a knife before the trim and molding are replaced.

Reflective or Metallic

Reflective or metallic insulation once was considered quite efficient but it fell from favor when it became apparent that it simply was not as good as other types. This insulation provides dead air spaces between two sheets of shiny metal, generally aluminum foil. Radiant heat, either from inside or outside the house is reflected back. The reflective material does bounce back radiant heat energy, but it also conducts heat away through conduction and even convection. The dead air spaces in this type insulation are quite large and, as a result, convection currents are created. The efficiency of any insulation is determined by spaces that are small enough so that any air contained in them does not move. When the spaces are large enough to allow convective currents, insulation efficiency drops off rapidly.

Foam Insulation

If you want foam insulation for your whole house, hire a professional. This installation requires professional equipment to mix the ingredients and to pump the mixture into wall cavities through holes that are cut in the wall between the studs. The holes are then later plugged. However, for the homeowner who has only a small area or space to fill, such as around a newly installed window or door, you can purchase small quantities of foam insulation. The foam fills irregular spaces as it expands and does a much more efficient job than batt, roll or loose insulation. For all its advantages, foam insulation is quite expensive and is suggested only for jobs where you cannot use other less expensive insulation materials. There have recently been cases where some foam insulation was found to give off toxic fumes if burned. Investigate whatever type of foam insulation you are considering. If necessary, get written guarantees as to its health safety.

Silicate Compound

Because of the present energy crisis and the need for efficient insulation, there has been considerable research on a variety of materials with insulation potential. One such material is a silicate compound, made of the same material as glass and sand. The silica is manufactured in such a way that the granules of material are puffed to create an insulation that will not burn, release toxic fumes or attract vermin. The product is odorless and non-irritating when poured into spaces in a home.

One of the newest types of insulation is made of silicate material. It has a higher R-factor than other loose insulation.

Sold under the trade name Dacotherm®, the insulation has an R-value of R-4 per inch as compared to vermiculite at R-2.2 and cellulose at R-3.6. The lightweight Dacotherm comes in bags. The homeowner opens the bag and pours the insulation. The BB-size pellets of insulation flow into every nook and cranny without dust or mess. Due to its low weight, the material can be handled by even senior citizens who might have trouble with heavier, more conventional pour-type insulations such as vermiculite. Still, it is a good idea to wear a mask and gloves.

R RATING

It has become common knowledge in recent years that a home needs from 6 to 12 inches of insulation. This is an oversimplification. You cannot select insulation just on the basis of thickness. Two kinds of insulation of the same thickness can have very different R-ratings. Check brands for R-rating, since the bigger the number, the better the results. Some labels will give the R-rating per inch of thickness; others will be on the overall thickness. Dividing the overall rating by the thickness will give the rating per inch, so that you can compare one insulation to the other. As with many things, generally the higher the R-rating the higher the cost. Conversely, the higher the R-rating, the less the heat loss through it.

INSULATION FOR WINTER HEATING

	Minimum	Maximum
Zone 1	—	R-19
Zone 2	R-11	R-38
Zone 3	R-11	R-49
Zone 4	R-19	R-57
Zone 5	R-19	R-66

The R-values are different for the walls, ceiling, and floor. The highest R-values are needed in the ceiling, the walls or floor, depending on the area of the country. In some northern areas, the frost line is so deep that the floors need more insulation than the walls.

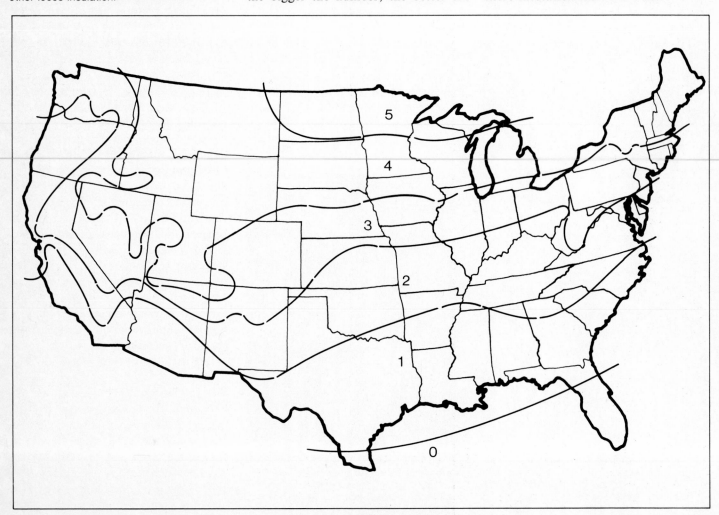

Heating zone map shows recommended R-values in different parts of the United States.

INSTALLING A VAPOR BARRIER ON MASONRY WALLS

Whether you are insulating a ceiling or a wall, there must be a vapor barrier between the surface and the insulation. The barrier keeps moisture away from the insulation, which could become so wet that the wood framing members could begin to rot.

There are a couple of ways to provide this very necessary vapor barrier. On masonry walls, you can apply one of the paints specially formulated to be a vapor barrier. A second way of providing a vapor barrier is to apply a layer of 4 to 6 mil polyethylene sheet plastic. This will work above ceilings, between studs and on the inside face of exterior masonry walls when you are going to finish the wall with paneling or wallboard.

To apply the film, use masonry nails or adhere it with construction adhesive applied with a caulking gun. Overlap the sheets of plastic 3 or 4 inches at the seams; staple the seams or seal them with adhesive. The polyethylene sheet plastic is sold in home centers and by larger mail order outlets such as Sears, Roebuck. The plastic is the same material recommended for laying over soil in a crawl space under a house to keep moisture from entering the crawl space.

If the wall is uneven, level it with furring strips, shimming as needed. Then you can staple the barrier to the furring.

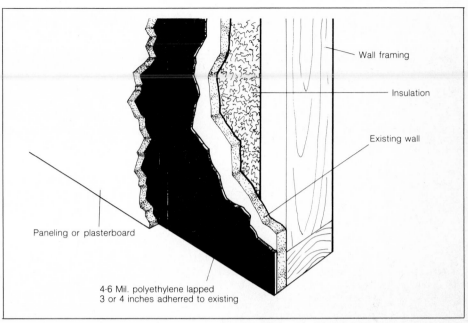

Paneling or plasterboard

Wall framing

Insulation

Existing wall

4-6 Mil. polyethylene lapped 3 or 4 inches adherred to existing

When insulation is added to exterior walls, a vapor barrier on the inside of walls is required. One method is to apply a sheet of 4 or 6 mil. polyethylene; then cover this with paneling or plasterboard.

INSTALLING INSULATION IN THE ATTIC

Because the largest amount of heat loss in winter, and heat gain in summer, is through the roof, the attic usually is the most heavily insulated area in your home. Unless living space is built in the attic, do not insulate the roof. Except for the southern states a thickness of 12 inches of insulation or more makes economic sense. This thickness offers an R-rating of about 30. In most areas walls should be insulated to at least an R-rating of 19, which means 5½ or 6 inches.

When insulating the attic always use a mask and wear gloves. If there are protruding nails in the ceiling, wear a protective head covering.

Over Existing Floor Insulation

It is not a good idea to add insulation to an attic in which there is poured or blown insulation installed but there is no vapor barrier. If you add insulation to this, you increase the chances of moisture condensing somewhere in the insulation and causing damage.

The best method is to remove the loose insulation and put down a vapor barrier. This could be a sheet of polyethylene placed down against the ceiling and brought up and over the joists. An alternate method is to remove the loose insulation, lay down roll or batt insulation with its vapor barrier down against

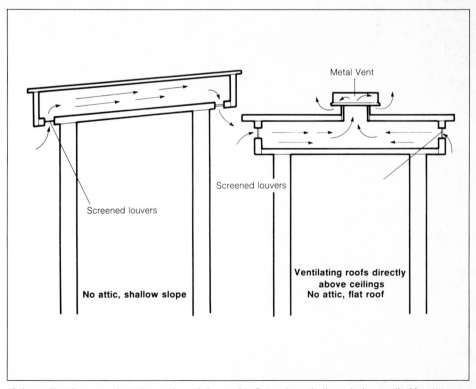

Metal Vent

Screened louvers

Screened louvers

No attic, shallow slope

**Ventilating roofs directly above ceilings
No attic, flat roof**

If the ceiling is up against the roof, or if the roof is flat or has shallow pitch, ventilation still must be provided.

The easiest way to install roll or batt insulation is to staple flanges to faces of wall studs or joists. However, it is not a recommended method.

The best way to install roll or batt insulation is to staple it to the inside faces of wall studs or joists. This assures an air space on both sides of the insulation.

the ceiling, and replace the loose insulation on top of the new.

Over An Existing Vapor Barrier If there already is a vapor barrier in an attic, do not install extra insulation that also has a vapor barrier. If you do, moisture could be trapped between the two vapor barriers and condense to wet the insulation. Wet insulation loses most of its insulating qualities.

If the ceiling joists in the attic floor are open, it is easy to pour loose insulation between them or to lay batt or roll insulation over the existing insulation. If the existing insulation is flush with the tops of the joists you can lay batt or roll insulation at right angles to the joists and existing insulation.

Insulating unfinished ceilings
Where there is no finished ceiling, batt insulation may be pushed between the joist spaces and secured to the bottom of the joists (batt insulation may be nailed or stapled). Batt insulation in 16- and 24-inch widths is available with R-values to 38 (for the coldest regions). If the insulation you are using does not come with a vapor barrier, staple a vapor barrier to the studs.

When installing the batts, fasten them under joist side, nailing through the flaps provided. Run the insulation over the plates at the exterior walls but take care not to block the eave vents, since

ventilation is a necessary component in insulation of a house or garage. Take care not to compress the batt insulation; this lowers its R-value.

Insulating Finished Ceilings
If Sheetrock is to be used on the ceiling, secure the batt insulation flaps between the joists to give the Sheetrock a tight fit (see illustration). If a vapor barrier does not come as part of the insulation, it will be necessary to add it. Cover the exposed faces of the joists (or studs) with the vapor barrier. Run the insulation over obstructions such as electrical boxes or light fixtures. However, if you have recessed light fixtures, insulation must be kept away from them (typically 3 inches away).

Insulating Around Obstacles
All insulation should go under wiring unless it compresses the insulation to do so. Start at the eaves with the insulation and work toward the center. It may be necessary to add light blocking at the eaves against which to butt the insulation without covering the eave vents. Be sure to fill around chimneys and other framed openings—unfaced insulation is best here. Insulate around openings such as pull-down stairs or scuttle holes. Glue the insulation in place around such movable accesses, for convenience in opening them.

INSTALLING INSULATION IN WALLS As a general note, do not leave faced insulation uncovered in any area where fire is a possibility; the facings are flammable. Cover insulation in such areas with suitable paneling; check with your building department.

Particular care should be taken when installing insulation in walls because of the narrow spaces and the many openings and obstructions: doors, windows, electrical service, etc. Faced insulation is easiest to use because of the nailing flaps provided at the edges.

If you are using a Sheetrock surface, you want a tight fit to the studs, so you should attach the batt insulation flaps between the studs.

Although batt and roll insulation often is installed by stapling the flanges to the

faces (narrow edges) of wall studs, this is incorrect. To be most efficient, there should be an air space on both sides of the insulation. Therefore, the flanges should be stapled to the wide sides of wall studs, rafters or joists, as shown.

The vapor barrier side of the insulation must always face the room. Secure barrier flaps to insides of studs. The flaps should be tight to the studs, with no gaps from which vapor can escape. If the vapor barrier is torn anywhere, tape it to repair it. Securing insulation between the studs leaves the stud faces bare: you may need to cover them also with a vapor barrier, depending on the type of construction and the locality; check with the building department.

Around Openings and Obstructions
Cover all doors and windows and other

obstructions. Then cut the openings out with a sharp knife, making sure the insulation around the obstructions is snug. Fill small gaps around obstructions with scrap insulation and cover these pieces with a vapor barrier. Check that the insulation fits snugly against the studs, sill, and plates that contain it.

If the wall material is not Sheetrock, the insulation may be attached to the face of the studs with nails or staples. Lap the flaps and keep them smooth to provide a smooth surface nailing. Unfaced installation follows the same procedures as above. Be sure the unfaced batts are snug between the studs and at the top and bottom. A polyethylene vapor barrier (4 to 6 mill is recommended) is easily installed. Secure it over all insulation and over obstructions and cut it out as you did the insulation.

INSTALLING INSULATION IN THE BASEMENT

One area where insulation often is ignored is in the basement. Any portion of a basement wall that is above the frost line will be the same temperature as the outside air. Below the frost line, the wall will be the average ground temperature, or 55 or 60 degrees (when dry) in most parts of the country.

In both cases, the temperature of the wall is below that of the basement air, which usually is at 68 or 70 degrees. The furnace can expend as much as 10 percent of its heat output attempting to warm the walls. Since heat always goes from the warm material to the cold, the basement wall soaks up heat constantly. An average home loses several million BTUs of heat energy through the basement walls every heating season.

Insulation Methods

You can insulate a basement wall in one of two ways. For the first, build a frame wall of studs against the concrete or concrete block wall. Staple roll or batt insulation to the framing. Then nail paneling or plasterboard over the insulation to the framing.

In the second method, adhere sheets of rigid insulation to the walls with construction adhesive. Apply ½-inch plasterboard over the rigid insulation as fire protection. Rigid insulation is made of a plastic, which burns with a dense smoke. Such a fire releases toxic gases.

Waterproofing Needs

Before insulating the basement walls, you must waterproof them. If you have the least suspicion that there might be water seepage through the wall, apply sheets of polyethylene plastic before building the framing for the inside walls. In this way any moisture that does seep through the wall will run down behind the plastic and emerge under the wall without damaging the framing or the wall covering. If you install rigid insulation, you need no vapor barrier. The board is waterproof. Do not neglect to use a waterproof adhesive.

INSULATING HOT WATER PIPES

Another item to insulate in the basement is the water piping. Hot water lines lose heat to the basement in the winter as well as the summer. Cold water lines will "sweat" in summer when humidity is high. The condensation can drip off the pipes and cause water damage.

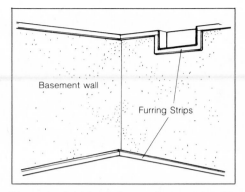

To apply rigid insulation on basement walls, first attach furring strips at windows and top and bottom.

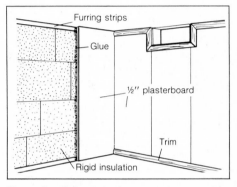

Then, glue ½ in. plasterboard to insulation with construction adhesive. Finally, add trim molding.

Insulation for pipes varies from material that spirals around the pipe on to neat sleeves that fit snugly around the pipe and pipe fittings.

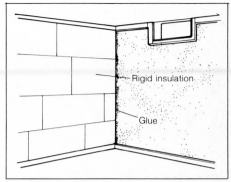

Cut sheets of rigid insulation to fit. Hold them to walls with the appropriate construction adhesive.

Insulation keeps hot pipes hot, cold pipes cold. Fit this shaped material around pipe; then squeeze on patented zipper-like closure to lock in place.

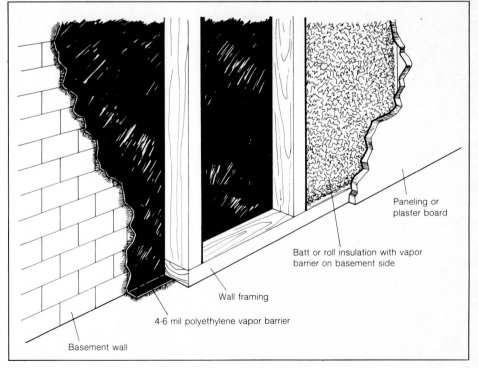

If there is a chance that there may be water seepage through basement wall, apply a sheet of polyethylene so water runs down behind it and exits at floor to avoid water damage to framing and paneling.

9

Humidity Control

Even before we all became energy conscious and made our homes as airtight as possible, controlling humidity was a problem.

WHAT IS HUMIDITY?
To better understand the problem, let's define "relative humidity." This is the amount of water vapor in the air at a given temperature, as compared to what that air would hold if it were completely saturated with water. A relative humidity of 50 percent, for example, means that there is half the possible volume of water vapor in the air. A relative humidity of 100 percent means the air is saturated with water vapor. At this point, condensation occurs.

Warm Vs. Cold Air
Warm air holds a greater volume of water vapor than cold air. This is why air that is cooled by an air conditioner loses moisture and you must provide a drain to handle the condensation. A dehumidifier works on the same principle; air is pulled through a condenser that lowers its temperature and causes the water vapor to condense out, either into a pan under the dehumidifier or into a drain hose.

The Effects of Humidity
On a hot, humid day you are uncomfortable because the air can accept no more moisture. As a result, there is little or no evaporation of moisture from your skin. On a day when the humidity is low, this moisture does evaporate, and you can be quite comfortable, even though the temperature is the same. Conversely, in the winter you will be comfortable at a low temperature if the humidity is around 40 percent. If the humidity is lower than 35 or 40 percent, there will be evaporation of perspiration from your skin and you will feel chilly.

The Ideal Situation
Controlling humidity, therefore, means that you provide enough water vapor in the air in the winter to keep human beings comfortable, yet not so much that it causes damage to the structure of the house. In the summer the condition is reversed; you will want low humidity to keep cool.

PAST SOLUTIONS
Uncontrolled Humidity
In the "old days," humid summer weather made doors and windows swell and stick so they were hard to open and close. People were uncomfortable as much from the high humidity as from the high temperatures. In the winter, inadequate insulation and weatherstripping caused the furnace to run most of the time. This produced very dry air. In turn, furniture and woodwork dried out, shrank and split. Windows and doors became so loose that they rattled in the wind. Because of the dry air, people often turned the thermostat up to 75 or 80 degrees to try to get warmer.

When we learned that the low humidity was a strong contributor to the problem, pans of water were placed near registers and behind radiators in an attempt to add moisture to the air. Unfortunately, these open containers of water did not do much to increase the humidity.

Added Humidity
The next step was to install powered humidifiers in central heating systems. Sometimes room-size units were used instead. Unfortunately, increasing the

Humidifiers can be installed in the ductwork of a forced air furnace, with the humidistat either in the ductwork or in a room upstairs.

humidity in cold weather created other problems that were more serious than physical discomfort.

Window Problems Windows "sweated" when the warm, moist air contacted their cold surfaces. The water condensed and ran down onto the windowsills to cause staining and even rotting. Storm sash helped reduce this condensation, but even then problems could arise. Either the window pane or the glass in the storm sash had condensation on the inside. The homeowner soon discovered that if condensation was on the inside window, then the storm sash was leaking cold air. This chilled the inner pane, and warm, moist air from the house condensed on the glass. If the outside sash had condensation on it, then the inside window was leaking, and

To prevent cold air from outdoors from entering space between windows use caulk to seal around storm sash and the framing.

Seal windows and storm sash that "sweat" with weatherstripping inside. Then warm moist interior air cannot enter between the sash.

warm, moist air from inside the house was passing into the space between the windows and condensing on the cold glass of the storm sash.

The seemingly obvious answer was to seal the window that leaked air. This sometimes worked. Weatherstripping around the inside window and/or caulking around the outside window stopped the condensation. Also, the house air felt warmer.

Wall Problems Not so obvious, but much more damaging in the long run, was the condensation that we did not

see, which had settled inside the exterior walls. This situation became apparent when outside walls were removed to remodel or add a room. Then the inside of the wall often was found to be saturated with moisture. The same situation that caused condensation on windows, created moisture on the inside surfaces of exterior walls, either inside the sheathing or the siding.

In homes where there was no insulation, the condensation settled on the inside of the wall and froze, if the temperature was low enough. Over weeks

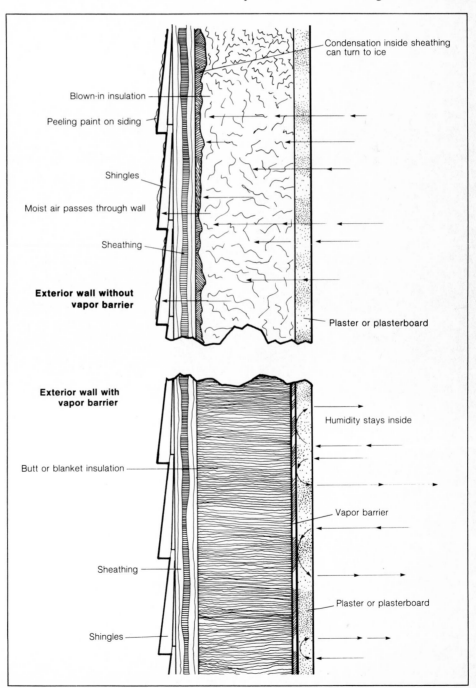

When there is no vapor barrier in an outside wall, moisture from the house condenses on the inside surface of the sheathing or siding.

or months of cold weather the frost built up inside the wall to a considerable thickness. When the weather warmed, the frost melted and water ran down inside the wall. After a few winters of constant wetness, the inside of the wall rotted.

Exterior Paint Peeling Even if houses have roll or batt insulation with a vapor barrier facing the interior wall, there usually are gaps in the vapor barrier that allow moisture to pass through to condense on the inside of the wall. If there is humidity inside the house, it will end up inside exterior walls.

If the paint is peeling on the outside of your house, one possible cause is that excess humidity is migrating through the wall and not only condensing inside the wall, but passing through the wood

of the siding and lifting off the paint. This condition also can indicate that you have no vapor barrier, especially if you have had cellulose insulation blown into the walls, or foam has been pumped into them. Neither of these has a vapor barrier, and moisture passes through the cellulose readily. Foam insulation offers some resistance to the passage of humidity, but there usually are areas in a wall where the foam does not fill completely and humidity can pass through.

The Ecology of Your Home

The several situations so far explained make it clear that each single thing you do to make your home more energy efficient affects, and is affected by, all the other steps you take. That is, weatherstripping seals the house and keeps in

the humidity. The high humidity can wet the insulation you install in the attic, walls or under the floor and reduce its insulating efficiency to almost zero. If you don't have a good vapor barrier in the outside walls, the moisture passing through the walls from inside can make the paint blister and peel, and so on.

Where Does Humidity Come From?

Humidity is the result of the various activities of the people living in a house. For example, washing and rinsing 80 sq. ft. of floor (an 8x10 room) produces almost 2½ pounds of water vapor. Drying 10 pounds of spun dried clothes creates 10 pounds of moisture, and cooking for a family of four results in almost 5 pounds of water vapor. An average shower, which equals four tub baths, creates from ¼ to ½ pound of moisture, and washing dishes for a family of four results in the release of 1 pound of moisture. Those same four people themselves release about 3 pounds of water vapor apiece during the course of a day. Gas appliances contribute 88 pounds of water vapor for each 1,000 cubic feet of gas burned. A humidifier (which can be controlled) will expell up to 48 pounds of water vapor into the air during a 24 hour period.

How Much Moisture is Necessary?

It is the consensus of authorities that inside humidity should not be higher than 40 percent in the winter. The following humidities were found to be optimum by the Engineering Experiment Station at the University of Minnesota, based on an inside temperature at 70 degrees F. If the outside air is minus 20 degrees or lower, humidity inside should not be over 15 percent. From minus 20 to minus 10 degrees, humidity should be not over 20 percent. From minus 10 degrees to 0 degrees, humidity should not be over 25 percent. From zero to 10 degrees, humidity should not exceed 30 percent, and from 10 degrees to 20 degrees, the humidity should not be more than 35 percent. From 20 degrees to 40 degrees the inside humidity should not exceed 40 percent.

However, if there is someone in your family with respiratory problems, the lower humidity levels can have negative effects. You may wish to supplement with a small one-room humidifier.

Overwatering house plants creates excess humidity. A moisture-checking device will show that they need less water than you think.

HOW TO SOLVE THE HUMIDITY PROBLEM

Eliminate excessive humidity and avoid expensive repairs by approaching the problem from four separate directions. First, avoid unnecessary evaporation. Second, discharge moist air to the outdoors, and preferably at the source of the water vapor. Third, ventilate so cool, dry air from outside enters the house. Fourth, provide protection to prevent condensation from occurring.

AVOIDING EXCESSIVE EVAPORATION

To avoid unnecessary evaporation, shut off all humidifiers, including the one in the ducting of the forced air furnace. When possible, hang laundry outdoors rather than in the basement or attic. Repair any faucets that are leaking; just a few drops a minute can result in gallons of water running down the drain, and evaporating on the way. Avoid overwatering plants, seal roof and wall leaks, and keep gutters clear. Provide a means of eliminating dampness from basements and crawl spaces. You also may need a dehumidifier.

DISCHARGING MOIST AIR

To implement the second step, make sure your clothes dryer is vented; this will stop condensation before it is created. All gas hot water heaters should be vented, as should gas ranges and ovens. In the past it has been common practice not to vent ranges and ovens, but with our new emphasis on making our houses airtight, the 88 pounds of moisture created for every 1,000 cubic feet of gas we burn cannot be tolerated.

Vent hoods over ranges sometimes are simply filters for grease and actually discharge the water vapor right back into the air in the kitchen. Replace this vent type hood with one that discharges out through the wall or roof. If this is not possible, investigate the new ranges that are vented down so that the ducts can be run down to the basement and out the wall.

Exhaust Fans

Laundries, kitchens and bathrooms all should have exhaust fans that are operated when moisture is created. Always close the doors to laundry rooms and bathrooms when water vapor is being created, and keep the doors closed until the exhaust fans have removed the water vapor.

Changing the Air

One way to increase ventilation in your home is to open windows slightly for a few minutes each day. Cool outdoor air contains less moisture than the warm air inside. When humidity in a home is especially high, open several windows and turn on exhaust fans to create a change of air in 15 or 20 minutes.

If you do plan to make an air change, balance the cost of heating the new air against the cost of repairing the damage that the humidity may cause.

Dampers

Humid air in a house has a "vapor pressure" that tends to move toward outside cooler, drier air, so opening the fireplace damper can aid in reducing high humidity. Do not forget to close it when the humidity has lowered, since a lot of heated air can go up a chimney. For a hot air furnace, a 3-inch duct can be run from outdoors to the cold air return of the system. A damper is installed to control the amount of air admitted. See Chapter 1 for instructions for running ductwork.

Room size dehumidifiers reduce the moisture level in one or two rooms, or a basement.

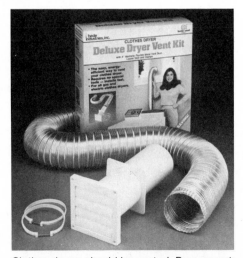

Clothes dryers should be vented. Proper venting will reduce drying time about 10 percent.

Opening the fireplace damper is one way to reduce humidity. Balance the loss of heated air against the need to reduce humidity.

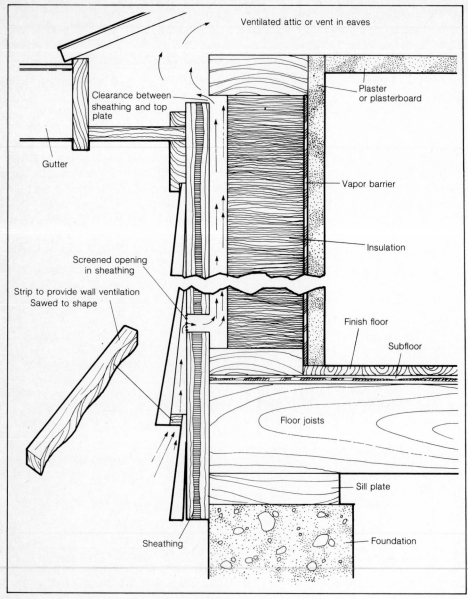

In extreme instances, ventilation is required inside a wall. Fit scalloped strips of wood under the bottom course of siding so air can pass through screened openings, rise up through wall and exit into a ventilated attic or out vents in soffits.

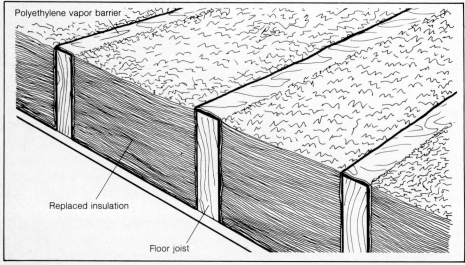

If existing insulation in attic does not have vapor barrier, remove the insulation, put down a vapor barrier, and replace the insulation. New insulation then can be added to increase the thickness.

Wall Circulation

There are instances when it is necessary to provide air circulation through walls so that moisture-laden air is removed. As indicated in the drawing, the ventilating air passes between the outside of the insulation and the exterior sheathing. The humid air is vented into an unheated attic that is well ventilated. As an alternative, the air can be vented directly outdoors through soffit vents.

Attic Circulation

As previously described, if a ceiling does not have a vapor barrier and the attic above is unheated, moist air will pass through the ceiling and insulation to condense on the underside of the ceiling planks. In the winter, the moisture can freeze. It later thaws, wets the insulation, and causes water stains on ceilings.

Moist air is difficult to predict; it will bypass a door and then go up through the walls as though they were chimney flues. Adequate ventilation in an attic is absolutely essential to protect the structure and to assure that insulation stays dry. Attic ventilation is discussed in detail in Chapter 3. Screened, louvered vents often are installed at opposite ends of a gable-roof. These vents are not efficient, however, unless one of them faces the prevailing wind. Without wind, the vents give no ventilation.

When room ceilings are directly under the roof rafters, as with some shallow roofs and "cathedral" ceilings, the spaces between the insulation and the roof decking should be ventilated from soffit vents that direct air to a continuous ridge vent or to other types of vents near the highest portion of the roof. If a roof is flat or only slightly slanting, the vent area should be 150th of the floor area, double that of more steeply sloped roofs.

A whole-house attic fan can also help remove moist air from the home. Some units come with both a thermostat and a humidistat. The latter switches on the fan when the humidity rises above a desired level (see Chapter 3).

PREVENTING CONDENSATION

The fourth measure is to provide protection in order to prevent the formation of condensation. This can be handled in several ways. Metal window frames, steel or aluminum, should be insulated to prevent cold from being conducted to the inside, which causes sweating. (In actuality, cold cannot be conducted as it does not exist. Cold is the absence of heat, and the metal frames conduct heat *away* so rapidly that the frames become cold.) This situation is not eliminated by storm sash that cover only the glass

area. Inside storm sash, such as those shown and described in Chapter 7, are a practical solution.

SPECIAL PROBLEMS
New Homes
If you move into a new home there will

be moisture contained in building materials, especially plaster, that is released very gradually for a year or two. To minimize this humidity, change the air frequently by using fans and opening windows. A dehumidifier will help eliminate the moisture more quickly.

HOW TO WATERPROOF A WALL

The best way to handle condensation in a wall space is to install a vapor barrier. You can apply an interior wall paint, such as "Insul-Aid" made by Glidden or similar products made by other paint manufacturers. As an alternative, you can strip off the wallboard or plaster, staple up sheets of 4 to 6 mil polyethylene plastic, then replace with a new wallboard installation. A big job, but much less expensive and time consuming than replacing the entire wall structure.

Another method of installing a vapor barrier is to adhere sheets of 4 to 6 mil polyethylene (overlapping seams 2 to 4 inches) to the existing wallboard or plaster. To attach the vapor barrier, either staple it to the wall studs or use adhesives designed for this purpose. Then apply paneling over the vapor barrier and wall covering.

Sealing Off the Basement
In some homes, the spaces between the floor joists on the first floor are not effectively sealed with a vapor barrier. This means that humid air from the basement or crawl space can move up inside the walls. These spaces should be sealed with sections of insulation plus a vapor barrier. This procedure also helps insulate the walls by preventing cold air from moving up inside them on the room side.

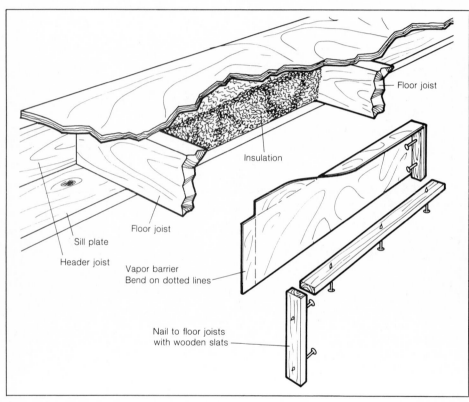

Floor joist

Insulation

Floor joist

Sill plate

Header joist

Vapor barrier
Bend on dotted lines

Nail to floor joists
with wooden slats

One way humidity from a basement or crawl space goes up into walls is through the open space between joists under the floor. Seal these with insulation backed by a vapor barrier.

Insulation has been stuffed into spaces at the ends of joists in this basement. Now a vapor barrier should be installed.

HOW TO TREAT A CONCRETE BLOCK HOME

In homes built in the 40s and 50s, you may find concrete block walls to which the interior plasterboard has been directly applied. This construction technique was much in favor when houses were built quickly to meet a tremendous demand. This type of construction provided no means of insulation, except by filling the cores of the concrete blocks (which often was not done) so there is a tremendous

and constant heat loss. Owners of such homes had a difficult problem keeping humidity low enough so it did not cause condensation on the inner surfaces of outside walls. Because of the construction method, condensation frequently occurred behind furniture set next to outer walls, because there was poor circulation of warmed air.

The only real solution to the problem is to build an auxiliary wall of 2x4 framing against the existing wall. Staple in-

sulation between the studs; then staple a vapor barrier of polyethylene over the insulation. Finally, nail wallboard to the framing. This construction arrangement does reduce the size of the room by 4 or 5 inches for each outside wall, and you must shim out the door and window frames. This procedure, however, is the only sure way to solve the insulation and vapor barrier problems presented by this type construction. It can also lengthen the life of your furnishings.

HOW TO TREAT A NEW BASEMENT

When a new house is built, or a basement is built for a room addition, you ensure a dry basement by applying a continuous layer of polyethylene sheeting on the outside of the foundation walls before backfilling. The same vapor barrier material should be placed on the ground before the basement floor is poured. The polyethylene comes in rolls 10 to 20 feet wide, and the strips applied to the foundation are overlapped 4 inches. If the basement is built in an area where there is a high water table, place tiling on the outside of the footing; lead a drain from the tile to a lower spot.

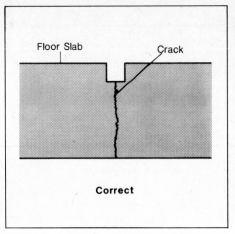

Chisel down to solid material before you attempt to repair a concrete crack.

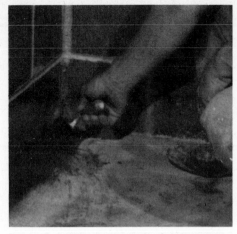

Seal all basement cracks with a recognized waterproofing material before coating the walls.

HOW TO TREAT AN EXISTING BASEMENT

As previously stated, moist air tends to move toward drier air, which can occur with air moving from a damp basement to the drier area upstairs. The simplest way to dry out a damp basement is to use a dehumidifier. This assumes that there are no structural causes for the humidity.

If there is water leakage in the basement, cracks must be sealed and then the walls and floor should be covered with waterproofing compound.

All downspouts should be directed into plastic or concrete splash pans that direct the rainwater away from the foundation. The basic fact is that ordinary concrete and concrete blocks, are not waterproof, so the best way to avoid water in a basement is to direct it away from the foundation walls.

PATCHING THE CRACKS

Clean out the crack with a cold or brick chisel, driving these tools with a baby sledge hammer. Make the cut in the form of an squared U, if you can. This will help hold the patch in place.

Press the hydraulic cement into the break and smooth it level with the surrounding surface of the foundation wall with a putty knife. You will have to work fast; hydraulic cement sets quickly. Wear gloves for this job, since hydraulic cement creates heat that could injure your skin.

WATERPROOFING A FOUNDATION

If the hydraulic cement does not work, and the rain-carrying (gutter) system is not the problem, there is only one sure way to stop the leakage: adding drain tile and waterproofing the foundation.

Step 1: Excavating Near Foundation Wall

Dig a trench around the foundation, wide enough for you to fit in it, and deep enough to reach under the foundation footing. Clean off the foundation wall; use a wide scraper for this. Scrub the wall down with water from a garden hose and a stiff broom or a brush.

Step 2: Laying the Drain Tile

Lay a 3-inch bed of medium-sized gravel in the trench. Then lay a row of field

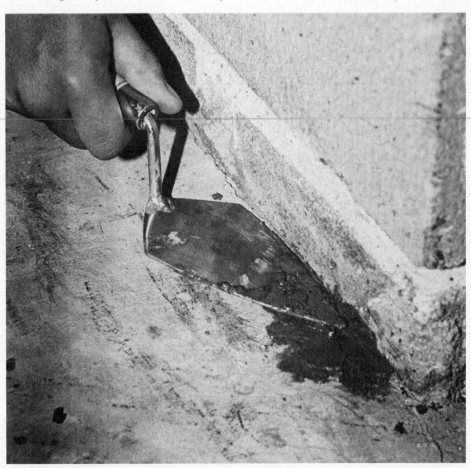

Press hydraulic cement into leaking foundation/basement floor joints. You can use your fingers to make the patch; then smooth with a trowel as shown.

tile on the gravel around the bottom of the foundation. The tile should have a slight pitch.

At one corner of the house, where the tile come together in the downward pitch, run a length of tile out into the lawn about 10 feet. You will need to dig a trench for this too. Lay the tile on 3 inches of gravel.

Step 3: Treating the Wall

Coat the foundation walls, from the footing to grade level, with a thick application of asphalt roofing cement. You can apply this with a trowel and brush. Make sure all areas are covered thoroughly. Embed a vapor barrier of black 4 mil polyethylene film into the asphalt roofing cement. Overlap the joints of the polyfilm about 4 inches, sticking the joints together with roofing cement. Then give all the vapor barrier a thin coating of the roofing cement.

Step 4: Finishing Up

Let the job dry for a couple of days.

Then backfill the earth into the trench. Make sure the fill slopes away from the house at a rate of about 1 inch per foot.

Save any leftover dirt. The ground will settle for some time and you will need this dirt to fill depressions.

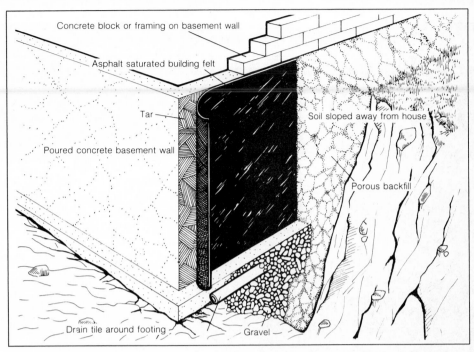

It is often worth your while to excavate a leaky basement and waterproof the outside walls.

HOW TO TREAT UNPROTECTED CRAWL SPACES Where the soil frequently is wet, crawl spaces are subject to excessive humidity that can migrate up through the floor into the house. Generous ventilation sometimes will get rid of the moisture-laden air. Vents in the foundation, located to take advantage of a prevailing wind, is one method. The area of all vents should be equal to 2 square feet for each 100 lineal feet of foundation perimeter plus one half percent of the crawl space area (see Chapter 3).

Frequently, foundation vents are not large enough to provide the needed ventilation, so an exhaust fan is installed. The lower end of the ducting is supported above the ground and the exhaust can be run to the outside through the floor joist or the header joist that runs at right angles to the floor joists.

Moisture Control

To control ground moisture, place a layer of the sheet plastic on leveled soil. Over this, smooth a 3-inch layer of sand. Waterproof the outside of the foundation wall and install drain tile around the footing to drain away any water that

might rise higher than the floor of the crawl space.

Floor Joists

An an alternative, the undersides of the floor joists can be sealed with a vapor barrier if water lines and ducts for heating do not make this impractical. If any

ducts or pipes do pass through the vapor barrier, they are sealed with duct tape to the barrier to assure a moisture tight surface. Seams in the sheet plastic are overlapped and stapled to the joists lengthwise, then further supported by nailing laths or other strips of wood to the joists over the seams.

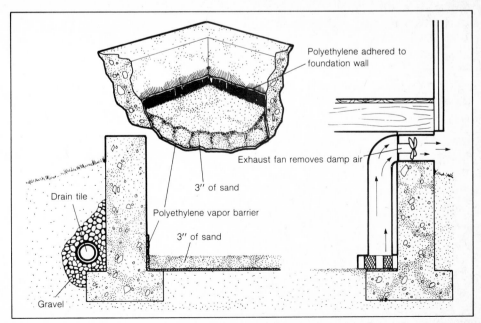

To waterproof a crawl space, provide a drain tile and slope the ground away from the foundation. Then, inside, install a vapor barrier and several inches of sand.

WHAT TO DO WHEN HUMIDITY IS TOO LOW

Up to this point we have talked about excess humidity; what if you don't have enough? This condition becomes apparent on a winter day when you walk across a carpet and feel a spark of static electricity jump from your finger to a light switch. When you comb your hair you can hear the sparks of static electricity, and you may even see them. You know the air is dry. When the humidity is near the optimum level, static electricity either does not occur or is drained off through the moisture in the air.

Geographical Factors

Some parts of the United States are more apt to have dry air than others. As the map from Sears, Roebuck shows, humidity requirements are highest in the states of Montana, Wyoming and Colorado and lowest in the Pacific coast and southeastern states. Numbers of the map indicate "average annual humidification hours," which are the hours in a typical year when the relative indoor humidity is less than the equivalent of 30 percent at 70 degrees F.

Humidifiers

You can use a central humidifier in the furnace duct system, although this can result in overkill. Better are single-room units that humidify only the areas in which there are people. The latter units will avoid increasing humidity in rooms, such as those with large window areas, where excess moisture can cause problems.

Clothes Dryer Bypass

One effective way to provide humidity, as well as "free" heat, to a home in the winter, is to install one of the several bypass valves now available for installation in the vent line of a clothes dryer. Even if the laundry room is in the basement, such a device can be useful, since most basements are quite dry in the winter when the furnace is running.

The filtered air from a dryer provides additional heat and humidity. This valve routes this air back into the house in cold weather.

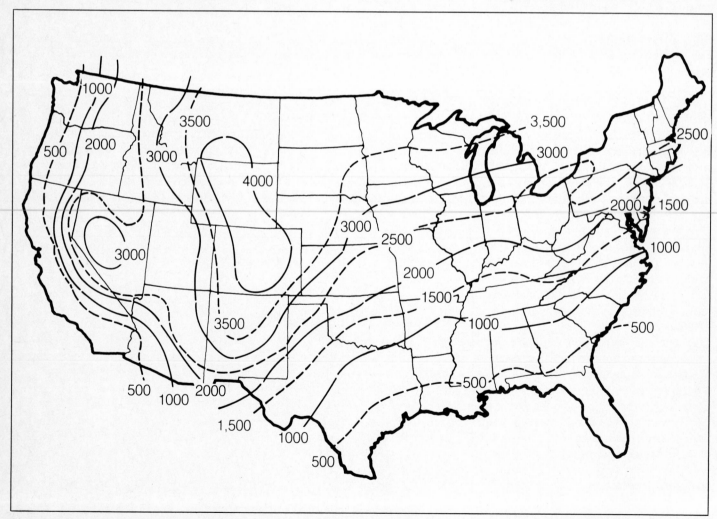

Map shows "humidification hours" as determined for the United States. The map clearly shows where excess humidity would be a problem and where lack of humidity could require added moisture.

10

Heating With A Wood Stove

In the last several years wood stoves have become more popular and have been considered by many people one way to offset the increasing cost of conventional fossil fuels. This may be true for some families, but for others the decision to heat with a stove can be an absolute disaster. Unless you have access to inexpensive and plentiful wood, a stove may be expensive. Unless you are prepared to be watchful and careful, the stove may be dangerous. A stove is also messier and dirtier than more modern heating methods.

COSTS AND CHARACTERISTICS

Stove prices vary enormously, from as little as $75 for a simple sheet metal unit to as much as $1,000 for a more exotic imported model. An efficient, reliable stove that meets UL standards will usually cost several hundred dollars. You will not save money by buying a really cheap stove because it will not be efficient and will consume much more wood than a quality stove. A stove must withstand a great deal of stress from the heat of the fire; a cheap stove may fail. Buy your stove from a reputable dealer and be sure that your stove is listed by the Underwriters' Laboratory or some other qualified testing agency.

However, not even the highest quality stove is safe unless it is properly installed. Proper installation costs money and may even require the construction of a new chimney. It is not safe (and against most building codes) to vent a wood stove into the same chimney as a conventional furnace.

Wood burning stoves produce a lot more particulates in the combustion process than do gas or oil fired stoves or furnaces. While this creates no real problem on an individual basis, if a large percentage of homes use wood burning heaters in a specific locality, air pollution becomes a real problem. In some towns and cities, such as Aspen, Colorado the use of wood burning stoves is prohibited by law because of an air pollution problem.

To combat this problem some manufacturers of wood burning stoves now include a catalytic converter that works much like the unit on the exhaust system of an automobile. The converter is a ceramic honeycomb coated with a precious metal that acts as a catalyst to burn the smoke and reduce polluting particulate by as much as 60%. It seems logical to assume that in the near future catalytic converters will be required on all wood burning stoves, and possibly even on coal burning stoves.

By the time you have added up the cost of the installation, including protection for the floor and walls (see "Safety Clearances"), the required metal smokepipe, insulating "thimbles" for the stack (smokepipe) where it enters the chimney and a container for carrying out ashes, as well as a poker for the fire and perhaps fire tongs, you can easily spend $1,000.

When you select a stove, consider that it will be a lifetime investment, not a disposable item you will replace in a few years. A quality wood burning stove should be self-amortizing; that is, it really should pay for itself.

Because of air pollution problems that accompany heavy use of wood stoves, at least one company has a stove with a catalytic unit that burns smoke and increases heat efficiency.

CHOOSING A HIGH-QUALITY WOOD STOVE

When it comes to buying a stove you have a choice of three basic material selections. Sheet metal is used for cheaper stoves. Sheet metal is the name for metal that is ⅛ inch or less in thickness. The metal pieces are bolted or welded together to create the stove. The better stoves are made of steel plate or cast iron. Stoves of plate steel are assembled from heavy sheets of steel from ¼- to ½-inch thick. The sheets are welded together. Doors, legs and other parts may be made of cast iron.

You can readily tell a cast iron stove from other types because patterns of various kinds will be cast into the several components to make the stove more attractive.

SIGNS OF QUALITY
Materials and Workmanship

The thicker the metal, the longer a stove should last. Because both cast iron and steel plate have similar characteristics, there is not too much difference between stoves made of these materials. Cast iron and steel plate retain about the same amount of heat per pound, and stoves made of either material cost about the same. Good workmanship, smooth welds and clean castings are the sign of a quality stove.

A 500-pound stove will continue to radiate heat for several hours after the fire has burned out. Some makes of stove will have firebrick or heavy metal plates inside to prevent burnout by the fire. The brick or metal also increases the thermal mass of the stove so it will store more heat for a longer time than a stove without such provisions.

Wood Stove Kits

Plans and kits of parts such as doors and legs are available to convert a 30 or 55 gallon oil drum into a wood burning stove. While these stoves are fine for a garage or camp, they are not too practical for heating a home. The sides of the drum can get red hot and present a hazard, especially to children and older people.

Efficiency Ratings

No matter what the material a stove is made from, it can vary considerably in design and efficiency. In the simplest types of stove, air moves in a straight path over or through the fire to assure combustion, but there is no provision for additional air for secondary burning. These lower efficiency stoves also have air leaks at the joints, which cause poor fire control because you cannot reduce the air supply to the point that the fire will burn slowly. Franklin stoves, pot belly stoves, box stoves and sheet metal stoves are included in this category. Efficiency runs 40 or 50% at best.

More sophisticated stoves are the medium efficiency types that have better control of both secondary and primary air supplies. Air leaks through seams and joints will be minor. Efficiency will run 50 or 60%.

High Efficiency Units The most technically sophisticated stoves are high efficiency units that regulate air flow like the simple airtight units, but additionally have baffles, long smoke paths and heavy heat exchangers that pull the most possible heat from the fire. These high efficiency wood burners are the most expensive to buy, but they cost the least to operate because they deliver the maximum heat per pound of wood. These high efficiency stoves, running from 60 to 65%, can be made of either plate steel or cast iron.

Using a Fireplace

While some homeowners consider fireplaces as an auxiliary heating plant, a typical masonry fireplace is only 10 or 15% efficient. Some even have a "negative" efficiency. In other words, more heat is drawn up the chimney by the fire than is delivered to the room. In fact, some of the heat already in the room is also blown up the chimney.

Much greater increase in efficiency of a fireplace is produced with a fireplace insert. Basically, this is a high efficiency wood burning stove that utilizes the fireplace chimney. You can fit a conventional wood burning stove into a fireplace in several ways. Install a tight fitting plate in place of the damper and run the stovepipe through the plate. As an alternative plan, you can place a covering over the fireplace opening and run the stovepipe run through the covering. This latter practice is not recommended, however, because there can be a build up of unburned gasses in the fireplace firebox that could cause an explosion. This would happen if the flue in the chimney were cold and a draft had not been started by having heat pass up through it.

FREE-STANDING FIREPLACES

If you want the beauty of a fireplace in combination with an efficient heating plant, install a free-standing fireplace or a Franklin stove. These units really are stoves, but each unit has provisions for opening the front so you can enjoy the flickering flames.

An efficient wood heater does not have to look old fashioned or inappropriate to modern furnishings. This unit, a heat-circulating fireplace, is well suited to this contemporary room.

OPERATING A WOOD BURNING STOVE

Wood stoves are much different than the automatic heating plants most of us have lived with all our lives. There is no thermostat to set that assures a constant temperature with no concern about the fuel supply. A wood burning stove needs constant attention. If you do not want this kind of bother, do not install a wood stove.

COMBUSTION PROCESS

In a furnace arrangement, you have no concern with the combustion process, but with a wood burning stove you must know the basics of building and maintaining a fire. Combustion requires a flame to start it, which in a wood burning stove is a match set to kindling wood, and fuel. Besides the heat and fuel you also need air. Control of this air supply determines how fast or how slow a fire burns, and the speed of burning determines how long a fire will last.

Stages of Combustion

There are three stages in the combustion process. First, heat drives off the moisture in the fuel. Second, at higher temperatures, "pyrolysis" breaks down the wood into gasses and coals. Third, at even higher temperatures the gasses and coals burn. All these steps may be taking place at once in a fire. Moisture will be driven from the center of the log, the outer portion will break down into gasses and coals, while at the top of the firebox the gasses and coals will be burning.

Damper Control for a Hotter Fire

It is the last reaction that is controlled by the "secondary" air supply in so-called "airtight" stoves. More simple stoves will have a damper in the chimney to control the flow of air through or over the fire. More complicated airtight stoves will have a "built in" damper that is controlled by a thermostat. When you adjust the thermostat for a hotter fire, the damper is opened to allow more air to enter the firebox. When you turn down the thermostat, the damper partially to allow less air to enter the firebox. As a result, combustion slows.

With experience, you will learn to control the air supply to the fire. Open-

Fireplace inserts reduce fireplace heat loss substantially. They burn wood in the same way as an airtight stove, increasing fireplace efficiency. Blowers transfer heated air to room.

A kit converts a 30 or 55 gallon oil drum into a stove useful only for cabins, garages and workshops because the shell gets red hot.

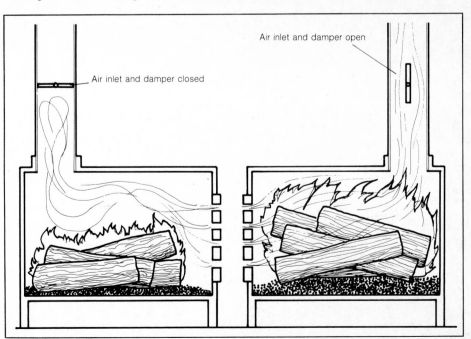

Air inlet and damper open

Air inlet and damper closed

The combustion rate and heat production in a stove is controlled by the use of an air inlet and damper. A low, long-lasting fire uses a smaller amount of air than a high fire.

ing the damper makes a hot fire, the room warms, and you are comfortable. But the fire quickly burns out and you need to add more wood. On the other hand, if the damper is closed too tightly, the combustion process is not complete. Then gasses produced by the pyrolysis stage in the fire go up the chimney without burning. You not only lose the potential heat of pyrolysis, but the cooler gasses tend to condense in the relatively cool chimney and settle out as creosote that will plug the chimney and present a fire hazard.

Starting the Fire

The very basic act of starting a fire can frustrate a beginning fire maker. If you learn the principles, however, the chore becomes easy. Check the instructions that come with your stove or follow these steps:

(1) open the damper and other air inlets on the stove;
(2) crumple newspaper and place it in the stove;
(3) stack kindling wood (small pieces up to 1 inch in diameter) on the paper in a pyramid shape;
(4) pile the kindling so air can circulate through the pieces;
(5) set a match to the paper;
(6) when the paper has ignited the kindling, and it is burning briskly, gradually add larger pieces of wood;
(7) after 10 or 15 minutes, close down the air controls to keep the fire at the desired level.

Safety Precautions Learn to start a fire with paper and kindling and never, but never take the shortcut of using gasoline or other flammable liquid to start a fire. A hot spot in the stove, or a few hot coals, can ignite the liquid and the stove may literally "blow up" with flames shooting out all openings. You, and your home, could get burned.

Maintaining a Fire

You have to learn how to make a fire last if you do not want to wake up to a cold house. The easiest method is to set your backup furnace at 55 or 60 degrees so it will come on and keep the house moderately comfortable if the fire in the wood stove burns out during the night. Or, if you have an early riser in your household, he or she can be given the task of stocking up the fire in the morning. Alternatively, someone can get up, stoke up the fire, then go back to sleep for another hour or so.

Thermal Momentum Another method for keeping the house warm sometimes is called "thermal momentum." You set the wood stove up so it overheats a room to 75 or 80 degrees. Even if the house cools down 15 or 20 degrees, the temperature will still be at an acceptable 55 or 60 degrees by the morning.

Low Air Supply If you have an "airtight" stove, you might be able to shut down the air supply so that the fire just barely burns. This is the method by which these stoves are said to hold a fire for 10 or 12 hours. However, if you do this very often you will have to clean the stovepipe and chimney frequently because the low fire will cause creosote deposits.

USING WOOD EFFICIENTLY
Types of Wood

Chunks of knotty wood that won't split with hammer and wedge will burn slowly all night, so save a few of these for just such use. Time your fires so they burn down to a bed of coals just before bedtime, so you can load up the stove with a full charge of wood. With a closed damper the fire will burn slowly all night. The same will work in the morning; load up the stove after shaking down the ashes and removing the excess. Then there will be a fire all day while you are at work. If someone will be home all day, be sure they know how to tend the fire and add more wood. Quite obviously, the hotter the fire, the more the wood that the stove will consume.

Green Wood While not efficient in every sense, mixing green wood with dry will cause the fire to burn more slowly. The temperature will not be as high as when using only dry, seasoned wood, but during the night or when you are away at work the lower temperature is acceptable.

Ash Disposal

One of the less pleasant chores associated with a wood burning stove is the removal of ashes. Use a metal shovel to clean out the ashes, and always put them in a metal container. It is always best to assume that there will be hot coals in the ashes. Never put them in a paper, cardboard or wooden container. This is asking for trouble. Do not dump the ashes in a garbage can, or you might have the odor of cooking garbage to remind you to put out a fire in the trash container. Ashes can be mixed with garden soil, and they also are handy for skidproofing driveways and sidewalks in the winter.

HEAT VALUE PER CORD
(In BTU per cord)[1]

High (24-31)	Medium (20-24)	Low (16-20)
Live oak	Holly	Black spruce
Shagbark hickory	Pond pine	Hemlock
Black locust	Nut pine	Catalpa
Dogwood	Loblolly pine	Red sider
Slash pine	Tamarack	Tulip poplar
Hop hornbean	Shortleaf pine	Red fir
Persimmon	Western larch	Sitka spruce
Shadbush	Juniper	Black willow
Apple	Paper birch	Large-tooth aspen
White oak	Red maple	Butternut
Honey locust	Cherry	Ponderosa pine
Black birch	American elm	Noble fir
Yew	Black gum	Redwood
Blue beech	Sycamore	Quaking aspen
Red oak	Gray birch	Sugar pine
Rock elm	Douglas fir	White pine
Sugar maple	Pitch pine	Balsam fir
American beech	Sassafras	Cottonwood
Yellow birch	Magnolia	Basswood
Longleaf pine	Red cedar	Western red cedar
White ash	Norway pine	Balsam poplar
Oregon ash	Bald cypress	White spruce
Black walnut	Chestnut	

USING A WOOD BURNING FURNACE

Quality wood burning furnaces and boilers can be connected to the ductwork or piping of your existing furnace or boiler. With this arrangement you can set the existing furnace to serve as a backup unit that comes on only when the fire in the wood burner goes out, or becomes so low it is not producing sufficient heat to warm the house.

Like a stove, a wood-burning furnace requires constant attention. There are not automatic systems for wood burning

furnaces, and if they ever are developed they probably will be expensive. In the past there were automatic stockers developed for coal burning furnaces, and it is possible that some such device may be created for wood. Even then, the furnace will demand frequent attention.

Efficiency A wood burning furnace generally will hold a fire for 12 hours, as its firebox is large and the air control is good. A thermostat can be used to control a wood burning furnace. This device will cut down the air supply when no heat is called for and then provide more air to make the fire blaze up when the thermostat calls for more heat. But the thermostat is useless unless you have loaded wood into the furnace and made sure the heat exchanger, stack and chimney are clean.

Wood Requirements A basic requirement for a wood burning furnace is a good supply of inexpensive wood, because an average unit will burn about 8 cords per full heating season.

Multi-Fuel Furnaces

If you are considering replacing an old furnace, you might take a look at the multi-fuel furnaces. These furnaces are fired with wood, but if the fire goes out or becomes low, the gas or oil burner comes on to boost the heat so the house stays comfortably warm. Some furnaces of this type have provision for starting the fire with the aid of the gas or oil burner.

Maintenance

No matter what setup you use with a wood burning or multi-fuel furnace, it will require more care and maintenance than a straight oil or gas unit. You must load in the wood, start the fire and add wood every once in a while. There also is the continuing job of removing the ashes and of cleaning creosote deposits from the heat exchanger, stack and chimney.

Local Codes

A wood burning or multi-fuel furnace may not be legal in all states because at the present time there are no code requirements to assure safe construction and maintenance of the units. Be sure to check the requirements for your particular area.

INSTALLING A WOOD STOVE SAFELY

More and more wood burning stoves are being installed in this country, and as a direct result there are more and more home fires caused by wood stoves that are improperly installed and/or maintained. Wood stoves are not inherently dangerous, but more than common sense is required for a safe installation and proper maintenance.

SAFETY CLEARANCES

If your wood burning stove is UL listed, follow the instructions packaged with the stove as pertains to clearances. Should you purchase a stove second hand or one that is not listed by UL or another testing laboratory, then follow the recommendations based on the standards of the National Fire Protection Association. The standards, based on many years of experience, provide reasonable (but not extreme) margins of safety, so do not reduce any of the clearances.

Side Clearance

The sides of the stove must be at least 36 inches away from any combustible surface. Do not reduce this spacing. Although high temperatures are required to set fire to most combustible materials, long exposure to heat will darken a material. The darker color will absorb more and more heat and will break down the material to the point where it will burn when exposed to temperatures as low as 200 degrees. This temperature is not at all unusual on an unprotected surface exposed to the heat.

Protective Materials

A stove can be placed closer than 36 inches from a wall if the wall is pro-

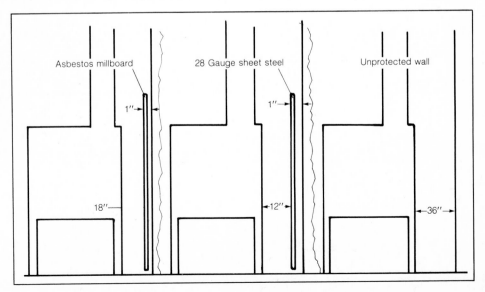

Asbestos millboard   28 Gauge sheet steel   Unprotected wall

1"   1"

18"   12"   36"

tected by a noncombustible material spaced a minimum of 1 inch away from the wall. This allows cooling air to circulate behind it. Do not place a sheet of noncombustible material directly on the wall, because the material will conduct the heat to the wall behind it and offers no protection at all. Asbestos millboard placed 1 inch from a wall will allow placement just 18 inches from the wall. A sheet of 28 gauge steel spaced 1 inch from the wall will permit installation as close as 12 inches.

Asbestos Millboard Because of some questions about asbestos being a possible carcinogen, be careful with asbestos millboard. It would be a good idea to paint the board with a high-temperature paint to reduce the chances of fibers breaking loose from the board. At the time this was written an effort was being made to find a fireproof substitute for the asbestos millboard, and one may now be available.

Local Codes Regardless of any clearances stated here, check your local building codes for the clearances required in your area. There have been so many fires caused by improperly installed wood burning stoves that some insurance companies are charging a higher premium where this kind of heating plant is used. Some fire insurance or homeowners' insurance could be voided by improper installation of a wood burning stove.

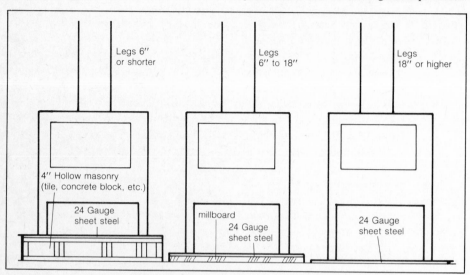

Combustible flooring must be protected from the heat of a wood stove. Stoves with short legs must be raised on a noncombustile platform topped with 24 gauge steel sheet.

Floor Clearance

Safe clearances between a stove and the floor are shorter than those between a stove and the surrounding walls because less heat generally is radiated from the bottom of the stove than from the sides. Ashes in the bottom of a stove have an insulating effect. Sometimes it is recommended that sand be placed in the bottom of a new stove to provide this insulation. The sand may not be necessary if the firebox comes equipped with firebrick, since the brick also acts as an insulator. Check the instructions with your stove to see how this should be handled for your particular unit.

Floor clearance is determined by the length of the stove legs. If the legs are 18 inches or higher a sheet of 24 gauge sheet steel is required on the floor. Legs 6 to 18 inches call for a sheet of 24 gauge steel over a layer of ¼ inch asbestos millboard or equivalent, while legs 6 inches or shorter require the 24 gauge steel placed over 4 inches of hollow masonry through which cooling air can circulate. Floor coverings can be made more attractive by using brick, stone or tile over them, but they cannot be used instead of the metal or asbestos sheet.

Front Clearance Sparks and embers falling or flying out the door of a wood stove present an additional safety problem, so the floor protection should be extended 18 inches in front of the stove and 6 inches along the sides and back. When removing ashes, make sure that any that spill, along with hot embers, fall only on the protected area of the floor.

Proper protection and standard clearances must be maintained, even in formal areas. The floor protection may be brick or tile to coordinate with the basic decor.

Furniture

Not only must a wood stove be kept clear of combustible materials, combustible materials must be kept away from a stove. Do not stack firewood close to a stove, and do not move furniture closer to it than 36 inches. A couch moved "temporarily" close to the fire could start to blaze after everyone has gone to bed, and the house could be destroyed. For the same reason, do not dry wet clothes on a wood burning stove without standing by and watching. A blazing mitten on an unattended stove could fall to the floor and set the house on fire.

Safety Provisions

Modern technology can help you keep your wood burning stove safe. Have a smoke detector in the room with the stove so that the detector will detect a fire while it is still a small smoldering hot spot that you can extinguish easily. A small fire extinguisher hung near the stove, in plain sight, can be used to put out a small fire before it becomes a big one. Be sure the extinguisher is near a door that would not be blocked by fire. An extinguisher is no good if you cannot reach it.

Finally, have a metal container for the ashes. There most certainly will be hot coals hidden in the ashes and the metal container will hold the coals safely. A container with a lid is even better; the lid will smother any potential fire before it starts.

STOVEPIPE AND CHIMNEY

Although the terms stovepipe and chimney are sometimes used interchangeably, the two are completely different. Do not use stovepipe as a chimney, since the thin metal can rust or corrode through quickly when exposed to the weather. Also, the thin metal has no insulating factor and will chill the combustion gases quickly. This will cause a buildup of creosote. A chimney fire caused by the creosote can burn through the metal stovepipe and set the house on fire.

Most chimneys are made of masonry or of fabricated units with double or triple walls that maintain higher temperatures than stovepipe. Therefore only a chimney can contain a chimney fire.

Stovepipe Specifications

Use stovepipe only to connect a stove to a chimney. Stovepipe should enter a chimney well above the height of the stove. Keep the run of pipe as short as possible. Have as few turns and bends as possible. The horizontal portion should be no more than three-fourths the length of the vertical part of the pipe and should rise at a minimum of ¼ inch per foot.

Pipe Sizes Different gauges of metal are used for stovepipe, and the pipe comes in different sizes. Buy the size that matches the outlet on your stove; purchase pipe that is no less than 24 gauge steel. Thicker is better; the heavier metal will last longer. Fit lengths of stovepipe tightly together and secure the joints with two or three sheet metal screws. The screws will help keep the joints from being shaken apart by the force of a chimney fire if one should occur.

Inspection Metal stovepipe, no matter the gauge, will last only two or three years. Inspect it frequently and replace any section that has rusted or corroded.

Damper Some kind of damper designed to control the flow of air to the fire is used on every stove. If a chimney fire occurs, it is important to be able to close the damper to shut off the air supply to the fire. Airtight stoves sometimes have a "built in" damper, but even non-airtight stoves should have a damper in the stovepipe.

Insulating Thimble Never run a stovepipe through a closet, attic or other enclosed area that is not open to constant visual inspection. A fire could burn through the stovepipe in such a location and start a fire. You would not be aware of the fire until it was well advanced.

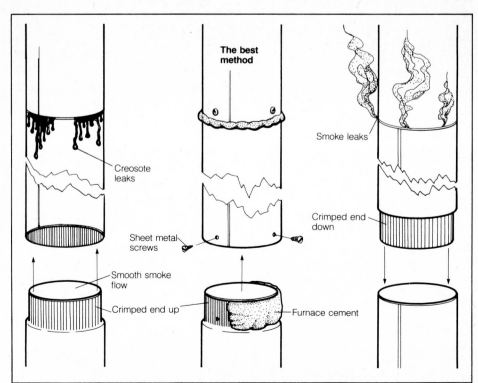

Overlapping stovepipe joint should cover the lower section with the upper to avoid smoke leaks. Furnace cement seals joint against creosote leaks for safest installation.

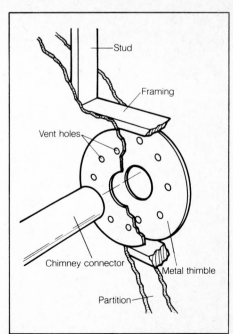

Fireproof thimble holds stovepipe passing through wall. Insulation protects the framing.

You should not route stovepipe through a wall, but if it is absolutely necessary, use a ventilated "thimble." Be sure the thimble is UL approved for wood burning stovepipe and is at least three times larger in diameter than the stovepipe.

A "thimble" also is used where a stovepipe joins to a chimney. It is permanently cemented into the chimney. The pipe must fit tightly in the thimble. The thimble and pipe must not extend into the chimney so as to reduce the flue size.

Chimneys

Chimneys play a critical role in the safety of your wood burning stove or furnace but they also present an unknown hazard. The main functions of a chimney are to create a draft that pulls air through the fire and to carry combustion products safely out of the house. Do not connect your stove to a flue to which another appliance is connected; each appliance should have its own flue. Carbon monoxide or smoke from the stove could escape from the unused appliance.

Dimensions To prevent backdrafts, do not use a chimney with a flue area more than twice the size of the stovepipe you are using for your stove. A chimney

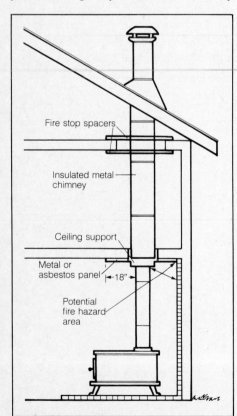

Even insulated stove pipes become very hot. Install extensive fire protection.

should extend at least 36 in. above flat roofs and be 24 in. higher than any point of the roof within 10 ft. This will prevent downdrafts and help keep sparks clear of the roof. If a chimney is not high enough to meet these requirements, have a mason extend the chimney and liner.

Checking an Old Chimney If you are going to connect your new woodstove to an old chimney, you must determine how safe the chimney is. A local building inspector or representative of the fire department will be glad to provide fire prevention assistance. Also, a mason who specializes in chimneys may be able to point out problems and tell you how he can correct them by tuckpointing the bricks or installing a new liner.

If you do not have access to experts, you can make some checks yourself. Look down the chimney to see if it has a liner. An old one should have a tile liner, which is a hard baked square or round clay tile. The tile will be much harder than the masonry; you can check this by scratching it with a knife. If the chimney does not have a liner, one must be installed and/or the chimney must be rebuilt completely. An alternate method is to install a metal liner that is designed for do-it-yourself installation.

One way to check a chimney for tightness is to build a smoking fire in the stove or fireplace. Then cover the top of the flue in the chimney. The cover will force smoke out any cracks in the chimney. Make this check quickly. Look in the attic and on all floors. A few friends to help would make the job go faster—and you will have less chance of filling the house with smoke. Make sure the friend on the roof can hear you when you shout up to tell him to uncover the flue.

Any leaks of smoke will indicate where the chimney needs repair. Make them as soon as possible and do not use the stove until the chimney has been made tight and safe.

Prefabricated Chimneys Because an existing chimney that is quite old would be expensive to rebuild or repair, many homeowners are buying prefabricated chimneys that have built-in insulation. These chimneys come in easy-to-handle sections that slip together.

Special fittings are available for running the prefab chimneys through walls and roofs. Be sure the prefab chimney you buy is UL listed for use with solid fuel appliances.

Masonry Chimneys Masonry chimneys are expensive to build, but they do have advantages that offset some of the cost. A chimney built inside a house, for example, has the ability to store an enormous amount of heat that later is released inside the house. The chimney can increase the efficiency of a heating system by up to 15% because of its thermal storage ability. In addition, because they are rust and corrosion resistant, masonry chimneys will outlast even the best metal prefabricated chimneys.

PROBLEMS WITH CREOSOTE

We have mentioned creosote several times in this chapter but have not fully explained what it is. Creosote is a black, tarry substance that is the result of condensation of unburned gasses of burning wood. When creosote accumulates in large quantities in stovepipes and chimneys, it can cause a chimney fire. The creosote is highly flammable, and a chimney fire will send a roaring column of fire 10 or more feet in the air with a sound like the roar of a jet engine. You cannot prevent the formation of creosote, but you can minimize it if you avoid smoldering fires, burn seasoned hardwoods and use inside chimneys that are not chilled by outside air.

Coping with Chimney Fires

If you ever do have a chimney fire, keep your head and do not panic. Immediately call the fire department, or have someone else call while you close all air inlets and dampers on the stove. Use a fire extinguisher on the stove to help smother the fire. At the same time, or immediately after using the extinguisher, wet down the roof to prevent sparks from setting it on fire. The fire department will take over when they get there.

Once the fire is out, go over your chimney brick by brick and joint by joint. The high temperature of the fire can crack the liner and blow out mortar joints. If the chimney is not repaired immediately, you may have another fire the minute you try to use the stove. Fix the chimney first.

CLEANING A CHIMNEY OR STOVEPIPE

Because of the formation of creosote, cleaning a chimney is an unavoidable chore when you use a wood burning stove or furnace. Frequent cleaning is best, but how frequently depends on the rate at which your stove burns the wood. This depends in part upon how well you maintain your fire.

CLEANING METHODS

One method of keeping chimneys and smokepipes clean is to build a very hot fire for a short time each day. This tends to clean out the chimney but does run the risk of causing a chimney fire. Never intentionally cause a chimney fire; it could damage the chimney and burn out the metal stovepipe.

Chemical cleaners can be used to break down the creosote deposits, but do not depend on the cleaners exclusively. Especially do not rely on them to remove heavy deposits; these must be cleaned out of chimneys and stovepipes by mechanical means.

At least once a year the chimney will require mechanical cleaning. You can hire professional chimney sweeps who will do the job for $50 and up per flue, or you can do it yourself.

CLEANING PROCEDURES

Step 1: Determining Buildup When you first start using a wood burning stove or furnace, check the flue every couple of weeks to determine the rate of creosote buildup. If the rate of creosote deposit is fairly slow, you can extend the time between inspections. At the start of a new heating season, or when a new batch of firewood is delivered, again go back to the two-week inspections until you determine what the rate of buildup under the new condition.

Step 2: Scheduling Inspections It is smarter to clean the chimney, and the stovepipe on a regular schedule than to wait until they are almost plugged with creosote. Then the cleaning job will be difficult and very dirty. And in the meantime the draft for the stove will have been reduced so that the stove's efficiency is way down. The lower the efficiency, the faster the creosote will build up in the pipe and chimney, so it makes good sense to clean frequently.

Depending on the stove or fireplace and the wood, the smokepipe and chimney may require cleaning once each month, but under no circumstances should you clean them less than once a year. It would be more prudent to clean the pipe and chimney twice a year even "if they do not need it." Preventive maintenance makes more sense than constant worry about a possible fire.

Step 3: Protecting Surroundings Before you begin, seal the opening inside the house with old sheets, pieces of cardboard or whatever is handy. If you want to clean from inside the house, make an opening in this enclosure, just large enough to accept the handle of the cleaning brush. Even then you can expect some mess in the house, so be prepared for it.

Step 4: Scrubbing the Chimney "Old fashioned" methods of cleaning a chimney such as dragging chains up and down, or lowering and raising a sack of rocks, are not as effective as using wire brushes designed specifically for cleaning chimneys. The brushes now are readily available in a variety of sizes and shapes to fit any chimney flue.

The modern chimney sweep uses the traditional brushes to clean soot and creosote from the chimney as well as more modern equipment such as a vacuum, safety goggles and a respirator.

THE QUESTION OF WOOD

Will you save enough on the cost of conventional fuel to pay for your wood burning stove? This depends on your source of firewood. A rule of thumb says that you should not consider a wood burning stove if you live farther than fifty miles from a source of wood. Even this does not always hold true in this day of ever increasing prices for everything. Wood is bulky and expensive to transport, and there are the further costs of cutting down the trees and splitting the resulting logs.

Price Range

As more and more people are buying wood burning stoves, more and more people are vying for the same wood supplies. If there are a lot of wood stove users in your area, the price of firewood is bound to be high. It is the old law of supply and demand. If the cost of wood is very high, a wood-burning stove may simply be an uneconomical investment.

Prices of firewood will vary with the locality and on transportation costs, as well as supply and demand. It usually is smart to buy in early spring when the demand for firewood is low. Stack your supply and season it for the following winter. Remember that the wood cut in spring when the demand for firewood is low. Stack your supply and season it for the following winter. Remember that the wood cut in spring may be wet and green, although if it has been cut in the winter it will have a lower moisture content.

Amounts of Wood

Cords Wood is sold in several standard units, of which the "cord" is best known. A standard for full cord should measure 4 feet wide by 4 feet high by 8 feet long so that its volume is 128 cubic feet. Because of the spaces between the irregular shaped pieces of wood, the actual volume of a cord of wood is more like 80 or 90 cubic feet. A dry cord of wood will weigh about 3,000 pounds, while a wet or unseasoned cord will weigh closer to two tons.

Face Cord A "face cord" is 4 feet high, 8 feet long and as wide as the lengths of wood are cut. These lengths can be 12, 16 or 24 inches to suit the size of your stove.

A face cord should cost less than a full cord because there is less wood, but this is not always the case, since you have to figure in the labor cost of the dealer who cuts the logs to your required length. If you buy 24 inch logs, you actually are buying just one half cord, while 16 inch logs make a third cord and 12 inch logs amount to only one fourth of a cord.

"Truck Load" Buying wood by the "truck load" is very risky. A standard half ton pickup truck will hold only from one third to one half a full cord. Never buy from a dealer who comes by in a pickup truck and says the load is "about a full cord." The best the truck could carry would be a half cord, and if the vehicle is a compact model, the bed would be too small for even that much wood.

Species of Wood

If you assume an average of 80 cubic feet of solid wood in a cord, there should be about 8,600 BTUs per pound of dry wood. This heating value is the characteristic you should consider most important when buying wood by species. Some types of wood are more dense than others and give off more heat when burned. These are the species of wood you should try to buy.

Mixed Lots Wood is characterized generally as hardwood or softwood, but there is a range of heating values among all types of wood and you must learn these values if you want to buy wood wisely. Most wood dealers, however, do not sort the wood according to species or grade, but cut and pile it as it falls. A cord of wood with a mix of grades is not a bad buy because it is convenient to have some lower quality softwood on hand to use as kindling and to make a quick fire for taking the chill off a room in spring or fall.

Heat Values The earlier table lists wood species by their relative heat values, figuring 80 cubic feet per cord and 8,600 BTUs for each pound of wood. The values are relative. Remember that wood will not ever be oven dry and produce 8,600 BTU/lb.

Moisture Content

The moisture content of wood is important because dry, seasoned wood will give you more heat per pound and ignite more readily. While wood that has been dried in an oven and burned under laboratory conditions will produce 8,600 BTUs per pound, wood never air-dries that well. The heating value of wet wood is reduced even more because some energy is required to first evaporate the moisture in the wood before the wood will burn.

Seasoned, air-dried wood may have a moisture content as low as 25 percent, but even this much requires about 4 percent of the heat to evaporate the moisture. You can figure that freshly cut, green wood will have a moisture content of 80 percent or more and 15 percent of the heat energy in the wood is lost just in evaporating away the moisture.

Seasoned Wood The term "seasoned" is rather vague, so if you buy seasoned wood try to learn how long it has been drying since it was cut. At least six months is required to properly dry wood for burning and many homeowners buy wood in early spring so it will dry all summer and part of fall. This assures that the wood is ready for burning and will produce the maximum BTUs per pound.

A bolted aluminum frame holds firewood neatly and keeps the wood accessible at all times.

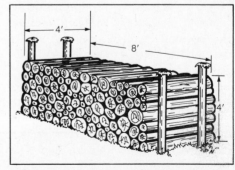

A standard cord of wood is 4 feet high, 8 feet long, and 4 feet deep. Any cord you buy should stack to this measurement.

SPLITTING YOUR OWN WOOD

If you want to save the most money when buying firewood actually cut down the tree, saw it into the proper lengths, and then split it. In this way, you do not pay for all the labor required to prepare the wood the way you want it. If you do not want to cut down the tree itself, shop around and buy wood in lengths from 4 to 16 feet. Have the logs delivered to your yard.

Muscle Power You must have a chain saw to cut the logs to the proper length and a means of splitting the shorter pieces. Start the split with an ax. Then hammer wood and metal wedges into the split. Drive a wedge into the split as far as the wedge will go. Then drive a second wedge alongside. The first wedge than will loosen. Remove it and drive it down alongside the second. Repeat this procedure until the log splits.

Log-splitting Devices You can purchase a powered log splitter. This may be run from an electric motor, a gasoline engine or even by a device bolted in place of the back wheel of your family car. Quality log splitters can run from several hundred dollars all the way to more than $1,000. This expense may be more than you want to bear in order to save money with your wood burning stove.

Time Required To cut, split and stack a cord of wood requires about 8 hours of labor. This goes along with the old saying "Wood warms you twice— once when you cut it, once when you burn it." You can figure that doing the physical labor pays a pretty good salary, but this could be questionable if you are not in good physical condition. Sawing and splitting a cord of wood is strenuous exercise and you'd better approach such a chore in slow stages to build up your muscles.

A chain saw makes firewood cutting a quick job. Log being cut must be securely supported.

Another modern innovation is a spring-powered maul. The spring increases the power applied and makes splitting easier.

BURNING CHARACTERISTICS OF WOOD

Hardwood	Heat Production	Ignition quality	Splitting quality	Smoke	Sparks	Comments
Apple, ash, beech, birch, dogwood, hard maple, hickory, locust, mesquite, oaks, Pacific madrone, pecan.	High	Difficult	Moderately easy	Light	If disturbed	High quality firewood
Adler, cherry, soft maple, walnut	Moderate	Moderate	Easy	Light	Few	Good quality firewood
Elm, gum, sycamore	Moderate	Moderate	Difficult	Moderate	Few	Good if well seasoned
Aspen, basswood, cottonwood, yellow poplar	Little	Easy	Easy	Moderate	Few	Good as kindling and first logs in fire
Softwood						
Douglas-fir, southern yellow pine	Moderate	Easy	Easy	Heavy	Few	Best of softwoods, but smoke a potential chimney problem
Cypress, redwood	Little	Easy	Easy	Moderate	Few	Usable
Eastern red cedar, western red cedar, white cedar	Little	Easy	Easy	Moderate	Heavy	Usable, best softwood kindling
Eastern white pine, ponderosa pine, sugar pine, western white pine, true firs	Little	Easy	Easy	Moderate	Few	Usable, good kindling
Larch, tamarack	Moderate	Easy	Easy	Heavy	Heavy	Usable
Spruce	Little	Easy	Easy	Moderate	Heavy	Usable, seasoned wood good kindling

Index

CONTRIBUTORS
ADDRESSES
PICTURE CREDITS

We wish to extend our thanks to the individuals, associations and manufacturers who generously provided information, photographs, line art, and project ideas for this book. Specific credit for individual photos, art and projects is given below with the names and addresses of the contributors. *Capital letters following page numbers indicate: T, top; B, bottom; L, left; R, right; C, center.*

Acorn Structures, Inc., P.O. Box 250, Concord, MA 01742 *(p.2)* **Appropriate Technology,** 14 Green St., Battleboro, VT 05301 *(p. 109 TL)*

Mr. and Mrs. Lawrence W. Babb, N52 W34293 Gietzen Dr., Okauchee, WI 53069 *(p. 5, 76, 77, 96)* **Bede Industries, Inc.,** 8327 Clinton Rd., Cleveland, OH 44144 *(p. 46 B, 113 BL)* **Berkus Group Architects,** 1531 Chapala St., Santa Barbara, CA 93101 *(p. 75 BL)* **Black Magic,** Box 977, Stowe, VT 05672 *(p. 135)* **Jim Blankets,** 240 S. Canon, Beverly Hills, CA 90212, *(p.6 BL, 7 BR, 78)* **Blueray Systems, Inc.,** 22 Berger St., Schuykill Haven, PA 17972 *(p. 14)* **Brick Institute of America,** 1750 Old Meadow Rd., McLean, VA 22101 *(p. 52, 69 R)* **Broan Manufacturing Co.,** Hartford, WI 53207 *(p. 45 TR)*

Carrier, P.O. Box 4808, Carrier Parkway, Syracuse, New York 13221 *(p. 25 TL)* **Certainteed,** P.O. Box 860, Valley Forge, Pennsylvania 19482 *(p. 112)* **Coleman,** 250 N. St. Francis, Wichita, Kansas 67201 *(p. 25 TR)* **Communicraft,** 3800 Regent St., Suite 3C, Madison, Wisconsin 53705 *(p. 53)* **Convenience Products, Inc.,** 4205 Forest Park Boulevard, St. Louis, Missouri 63108 *(p. 113 BC)*

Deal Products, Inc., 27th & Dearborn Street, P.O. Box 667, Easton, Pennsylvania 18042 **Diamond Shamrock Corporation,** Industrial Chemicals Unit 351 Phelps Court, P.O. Box 2300, Irving, Texas 75061 *(p. 114)* **Dow Chemical U.S.A.,** 2020 Dow Center, Midland, Michigan 48640 *(p. 113 CL, 117)*

EGO Productions, James E. Auer, 1849 N. 72nd Street, Wauwatosa, Wisconsin 53213 *(p. 5, 76, 77, 96)* **Enderes Tool Co.,** 924 14th Street, Albert Lee, Minnesota 56007 *(p. 129 BL)* **Emerson-Chrolan,** 8100 W. Florissant Avenue, St. Louis, Missouri 63136 *(p. 12, 44 B, 45 TL)* **EPRI-Electric Power Research Institute,** 3412 Hillview Ave., Palo Alto, California 94304 *(p. 51 T)*

Four-Seasons Greenhouses, 910 Route 110, Farmingdale, New York 11735 *(p. 83 L)* **Franklin Chemical Industries, Inc.,** 2020 Bruck Street, P.O. Box 07802, Columbus, Ohio 43207 *(p. 127)*

Garden Way Solar Greenhouses, Charlotte, Vermont 05445 *(p. 81)* **General Electric Major Appliance Business Group,** Appliance Park, Louisville, Kentucky 40225 *(p. 33 L)* **Frank Gehry, FAIA,** Architect, Santa Monica, California *(p. 50)* **Green Horizons,** Rt. 7, Box 124 MS, Santa Fe, New Mexico 87501 *(p. 6 TL, 7 TR, 73, 74, 75 T)* **Geocel Limited,** P.O. Box 653, Elkhart, Indiana 46515 *(p. 119 TL)*

Habitat Specialty Buildings, 123 Elm Street, S. Deerfield, Massachusetts 01373 *(p. 82 T)* **Heat Handlers,** P.O. Box 105, Addison, Illinois 60101 *(p. 126 CR)* **Homelite Division, Textron Inc.,** 14401 Carowinds Boulevard, Charlotte, North Carolina 28217 *(p. 39)* **Hunter Fans,** Hunter Division, Robbins and Myers, Inc., P.O. Box 14775, Memphis, Tennessee 38114 *(p. 41 T, 49, 143)*

Janco Greenhouses, J.A. Nearing Co., Inc., 9390 Davis Ave., Laurel, Maryland 20810 *(p. 82 B, 83 R)* **Johns Manville Corporation,** P.O. Box 5108, Denver, Colorado 80217 *(p. 116)*

K-Lux, K.S.H. Inc., 10091 Manchester Road, St. Louis, Missouri 63122 *(p. 107 RC, B)* **Gary Kucko,** 110½ N. Main Street, Rice Lake, Wisconsin 54868 *(p. 92-95)*

Lennox Industries, Inc., 200 S. 12th Avenue, Marshalltown, Iowa 50158 *(p. 13)* **George Leone,** photographer *(p. 59 R)* **Lord & Burnham, Costich & McConnell, Inc.,** 225 Marcus Boulevard, Hauppauge, New York 11787 *(p. 72 B, 75 BL)*

McGraw Edison Air Comfort Division, 704 N. Clark, Albion, Michigan 49224 *(p. 27, 30 L, C)* **Macklinberg-Duncan Company,** 4041 N. Sante Fe, Oklahoma City, Oklahoma 73118 *(p. 119 BL)* **Martin Industries,** 301 E. Tennessee Street, Florence, Alabama 35630 *(p. 121 TL, 128)* **Marvin Windows,** Warroad, Minnesota 56763

National Greenhouse, P.O. Box 100, Dana, Iowa 62557 *(p. 84 T)* **E.A. Nord Co.,** Everett, Washington 98206 *(p. 120)* **Richard V. Nunn,** Media Mark Productions, Falls Church Inn, 6633 Arlington Boulevard, Falls Church, Virginia 22045 *(p. 15)* **Nu-Sash,** 324 Wooster Road, N. Barberton, Ohio 44203 *(p. 105 L)* **Nutone,** Madison and Red Bank Roads, Cincinnati, Ohio 45227 *(p. 40 B, 41 B, 43, 46 T)*

Plaskolite, P.O. Box 1497, Columbus Ohio 43216 *(p. 107 CC, 108 TL, TC)*

Quaker Manufacturing Company, 705 Chester Pike, Sharon Hill, Pennsylvania 19079 *(p. 98)*

Ready-Built Products, P.O. Box 4306, Baltimore, Maryland 21223 *(p. 121 BC)* **Reynolds Metals Company,** P.O. Box 27003, Richmond, Virginia 23261 *(p. 105 BC, 106, 111 BC, 136)*

Sears Roebuck, Sears Tower, Chicago, Illinois 60606 *(p. 25 BL, 126)* **Sloan Valve INTERburner,** 10500 Seymour, Franklin Park, Illinois 60131 **Solar Vent,** 1091 Shary Cr., Concord, California 94518 *(p. 84 BR)* **Solar Age Magazine,** Solar Engineering Co., Westlake Village, California 91361 *(p. 6 R, 51 BL, 59 R)* **Solar Usage Now, Inc.,** 450 Tiffin Street, Bascom, Ohio 44809 *(p. 63 TR, CR)* **Stonelite Tile Company,** 1985 Sampson Avenue, Corona, California 91720 **Tim Street-Porter,** 6938 Camrose Drive, Los Angeles, California 90068 *(p. 50)* **Suburban Manufacturing Company,** North Broadway Street, Dayton, Indiana 37321 *(p. 34 R, 129 TR, 131)* **Sun-Day Solar,** 2701 S. 29th St., Milwaukee, Wisconsin 53215 *(p. 58 B)* **Sun Quilt,** Box 374, Newport, New Hampshire 03773 *(p. 108 B)*

Tempvent, Hwy 150 E, Shelby, North Carolina 28150 *(p. 46 T, 47)* **Thermal 81 Energy Rods, Pipe Systems, Inc.,** 1533 Fenpur Drive, Fenton, Missouri 63026 *(p. 59 L)* **Thermal Technology of Aspen Inc.,** 600 Alter Street, Broomfield, Colorado 80020 *(p. 109, 110)* **Thermon Manufacturing Company,** Thermon Drive, San Marcos, Texas 78666 **Thermo-Control, National Stove Works,** Howe Caverns Road, Cobleskill, New York 12043 *(p. 8 BR)* **Thoro Systems Products,** Division of Standard Dry Wall Products, 7800 NW 36th Street, Miami, Florida 33166 *(p. 124)* **Tile Institute,** c/o Lis King, Box 503, Mahwah, New Jersey 07430 *(p. 69 L)*

Vegetable Factory Greenhouses, 100 Court Street, Copiague, New York 11726 *(p. 72 T)* **Versol U.S.A.,** 252 Nassau St., Princeton, New Jersey 08540 *(p. 111 BR)*

Waterford Stoves, c/o Capitol Export Corporation, 8825 Page Boulevard, St. Louis, Missouri 63114 *(p. 132)* **Window Quilt,** A.T.C. Box 975, Battleboro, Vermont 05301 *(p. 109 BL)*

Glossary

Absorption and reflection These two factors account for the loss of solar radiation in the atmosphere. It is absorbed by the atmosphere or reflected by cloud cover. The actual amount of radiation that an area receives depends upon its location.

Active solar system A system that requires another energy source in addition to the sun in order to function. A pump, for instance, is incorporated into an active solar system in order to move heated water to a storage area.

Conduction Transfer of heat by movement from particle to particle. Heat passes through an aluminum storm window frame by conduction. A thermal break, usually vinyl, prevents conduction from occurring.

Convection Transfer of heat by the movement of a fluid or of air. Moving air removes heat from a collector's surface. Glazing keeps the moving air away, minimizing convection.

Cord Measure of wood. A cord is 4 feet wide by 4 feet high by 8 feet long.

Creosote Black, tarry, highly flammable residue that coats the inside of a stovepipe or chimney. You must clean a chimney of creosote at least once every year.

Damper Metal flaps inside ductwork and chimneys. Dampers open and close to regulate the amount of air that flows through.

Diffuse radiation Clouds and particles in the air absorb, reflect and scatter sunlight. As a result, only part of the sun's radiation reaches the earth. The scattered light is called diffuse radiation. At best, diffused radiation can only deliver about one quarter of the solar constant (see "Solar Constant" below).

Double shell and tube heat exchanger This system can be used to heat drinking water. Basically, the shell in this case has two types of compartments. Into one flows the toxic heat transfer medium. In the other is a nontoxic substance. The toxic substance's heat is transfered to the nontoxic substance. The tube runs through the nontoxic substance and the water inside the tube absorbs the heat.

Double wall heat exchanger This system also is appropriate for drinking water. In this case, the tube spirals around the outside of a tank containing the toxic heat substance. The heat passes through the walls of the tank and the tubing in order to warm the water.

EER Energy Efficiency Rating. A figure that indicates the amount of electricity an appliance needs in order to operate. The higher the EER, the more efficient the machine and the less cost to run.

Face cord A volume of wood that is 4 feet high by 8 feet long, but is cut to a specific width for a particular stove. The widths can run 12, 16 or 24 inches.

Greenhouse effect The trapping of energy behind glazing, because glazing permits short wave solar radiation to pass through to an absorbant surface. However, this energy reradiates as long wave energy. Glazing does not permit much of the long wave energy to pass back out through the glass. The long wave energy becomes bottled inside, producing the greenhouse effect.

Gross vent area Size of vent area required when the obstruction from screening and/or louvers is taken into account.

Heat exchanger A device designed to transfer heat from one fluid to another. In some solar heating systems a heat exchanger is required between the substance sent through the collector and the substances used for storage, or between the storage substance and the distribution substance. A system using antifreeze, for instance, needs a heat exchanger in order to provide potable water for household use. There are three basic types of heat exchangers: shell and tube, double shell and tube, and double wall.

Insulating thimble Accepted device for running a stovepipe through a wall or into a chimney.

Natural ventilation Air circulation throughout the house. Relies on non-powered devices such as vents or turbine ventilators. Cool air enters lower vents, is warmed, and rises to exit out upper vents.

Net free vent area Clear unobstructed open vent space required by a given floor space or a given powered fan.

Orientation A collector can sit within 20 degrees on either side of true south. Such factors as the design of the collector and the presence of shade will determine whether the angle is to the west or the east.

Powered ventilation Air circulation that is enhanced by mechanical fans mounted in the roof, attic or wall. In order to work correctly and quietly, power ventilation requires sufficient vent area.

Pyrolysis The second stage of a wood fire. During pyrolysis, the wood breaks down into gasses and coals.

Passive solar system A system in which no other energy source is used to activate the system. For example, heat absorbed by a Trombe wall is reradiated into a room by natural law, not by a fan or other mechanical means.

Sacrificial anode When the tubing in a solar collector is not of the same material as the plumbing lines, a galvanic chemical reaction takes place that can ultimately destroy the tubing. To prevent this from happening, a material that is expendable—such as a piece of screening—is inserted in a section of pipe. This material becomes affected rather than the other metals in the system. After about three years you will have to replace the anode.

Shell and tube heat exchanger This system consists of an outer shell, which encloses a toxic heat transfer fluid such as antifreeze, and a tube that runs through the shell. The tube carries potable water. As the water passes through the tube it absorbs heat from the transfer fluid. This water can be used for heating but not for drinking. This is because the tube could spring a leak and absorb some of the toxic substance.

Solar constant A measure of the solar radiation that strikes the outer atmosphere, 429.9 BTUs per square foot per hour. About half of this energy is lost in the atmosphere before it reaches the earth's surface.

Thermal pond A passive system that relies on roof-mounted stored water in order to provide thermal storage. The stored heat then radiates downward into the living space. In order to function properly, there must be provisions for protecting the storage bed from the air when the sun is not shining. Usually, this protection takes the form of insulating panels that slide into place over the storage area as needed. Another way to control the process is to remove and refill the water according to need. In some cases, the roof area is covered with a transparent roof structure.

Thermosiphon Natural movement of a fluid or of air. Hot air or water rises. A passive system utilizing thermosiphoning will draw cold water in below a collector. As it is heated, it rises through the collector. It is stored in a tank near the top of the collector until it is used or drawn into an auxiliary heater.

Tilt A solar collector is most efficient when the sun's rays fall upon it at a right angle. To determine the best angle for a collector, find the degrees latitude of your home. To this, add 10 degrees to 15 degrees—or subtract 10 degrees to 15 degrees. This will give you the range of angles from which you can select the angle of your collector.

Trombe wall Masonry or concrete wall designed specifically for thermal storage. A Trombe wall must be behind an expanse of glazing to function effectively.

Vapor barrier Sheet of 4 to 6 mil polyethylene used to prevent moisture seepage through such things as walls, cement slabs, and ceilings.

Metric Conversions

LUMBER

Sizes: Metric cross-sections are so close to their nearest Imperial sizes, as noted below, that for most purposes they may be considered equivalents.

Lengths: Metric lengths are based on a 300mm module which is slightly shorter in length than an Imperial foot. It will therefore be important to check your requirements accurately to the nearest inch and consult the table below to find the metric length required.

Areas: The metric area is a square metre. Use the following conversion factors when converting from Imperial data: 100 sq. feet = 9.290 sq. metres.

METRIC SIZES SHOWN BESIDE NEAREST IMPERIAL EQUIVALENT

mm	Inches	mm	Inches
16 x 75	⅝ x 3	44 x 150	1¾ x 6
16 x 100	⅝ x 4	44 x 175	1¾ x 7
16 x 125	⅝ x 5	44 x 200	1¾ x 8
16 x 150	⅝ x 6	44 x 225	1¾ x 9
19 x 75	¾ x 3	44 x 250	1¾ x 10
19 x 100	¾ x 4	44 x 300	1¾ x 12
19 x 125	¾ x 5	50 x 75	2 x 3
19 x 150	¾ x 6	50 x 100	2 x 4
22 x 75	⅞ x 3	50 x 125	2 x 5
22 x 100	⅞ x 4	50 x 150	2 x 6
22 x 125	⅞ x 5	50 x 175	2 x 7
22 x 150	⅞ x 6	50 x 200	2 x 8
25 x 75	1 x 3	50 x 225	2 x 9
25 x 100	1 x 4	50 x 250	2 x 10
25 x 125	1 x 5	50 x 300	2 x 12
25 x 150	1 x 6	63 x 100	2½ x 4
25 x 175	1 x 7	63 x 125	2½ x 5
25 x 200	1 x 8	63 x 150	2½ x 6
25 x 225	1 x 9	63 x 175	2½ x 7
25 x 250	1 x 10	63 x 200	2½ x 8
25 x 300	1 x 12	63 x 225	2½ x 9
32 x 75	1¼ x 3	75 x 100	3 x 4
32 x 100	1¼ x 4	75 x 125	3 x 5
32 x 125	1¼ x 5	75 x 150	3 x 6
32 x 150	1¼ x 6	75 x 175	3 x 7
32 x 175	1¼ x 7	75 x 200	3 x 8
32 x 200	1¼ x 8	75 x 225	3 x 9
32 x 225	1¼ x 9	75 x 250	3 x 10
32 x 250	1¼ x 10	75 x 300	3 x 12
32 x 300	1¼ x 12	100 x 100	4 x 4
38 x 75	1½ x 3	100 x 150	4 x 6
38 x 100	1½ x 4	100 x 200	4 x 8
38 x 125	1½ x 5	100 x 250	4 x 10
38 x 150	1½ x 6	100 x 300	4 x 12
38 x 175	1½ x 7	150 x 150	6 x 6
38 x 200	1½ x 8	150 x 200	6 x 8
38 x 225	1½ x 9	150 x 300	6 x 12
44 x 75	1¾ x 3	200 x 200	8 x 8
44 x 100	1¾ x 4	250 x 250	10 x 10
44 x 125	1¾ x 5	300 x 300	12 x 12

METRIC LENGTHS

Lengths Metres	Equiv. Ft. & Inches
1.8m	5' 10⅞"
2.1m	6' 10⅝"
2.4m	7' 10½"
2.7m	8' 10¼"
3.0m	9' 10⅛"
3.3m	10' 9⅞"
3.6m	11' 9¾"
3.9m	12' 9½"
4.2m	13' 9⅜"
4.5m	14' 9⅓"
4.8m	15' 9"
5.1m	16' 8¾"
5.4m	17' 8⅝"
5.7m	18' 8⅜"
6.0m	19' 8¼"
6.3m	20' 8"
6.6m	21' 7⅞"
6.9m	22' 7⅝"
7.2m	23' 7½"
7.5m	24' 7¼"
7.8m	25' 7⅛"

All the dimensions are based on 1 inch = 25 mm.

NOMINAL SIZE (This is what you order.)	ACTUAL SIZE (This is what you get.)
Inches	Inches
1 x 1	¾ x ¾
1 x 2	¾ x 1½
1 x 3	¾ x 2½
1 x 4	¾ x 3½
1 x 6	¾ x 5½
1 x 8	¾ x 7¼
1 x 10	¾ x 9¼
1 x 12	¾ x 11¼
2 x 2	1¾ x 1¾
2 x 3	1½ x 2½
2 x 4	1½ x 3½
2 x 6	1½ x 5½
2 x 8	1½ x 7¼
2 x 10	1½ x 9¼
2 x 12	1½ x 11¼